Current Topics in Neurotoxicity

Volume 12

Series editors

Richard M. Kostrzewa, Johnson City, USA
Trevor Archer, Gothenburg, Sweden

Current Topics in Neurotoxicity consists of volumes that cover the latest developments in neurotoxicity studies including neurodegeneration, neuroprotection, and neuroregeneration. The series has a translational focus and addresses the causes of neurotoxicity associated with diseases such as Alzheimers, Parkinsons Disease, autism, HIV associated dementia etc.

More information about this series at http://www.springer.com/series/8791

Jolanta B. Zawilska
Editor

Synthetic Cathinones

Novel Addictive and Stimulatory Psychoactive Substances

Springer

Editor
Jolanta B. Zawilska
Department of Pharmacodynamics
Medical University of Łódź
Łódź
Poland

ISSN 2363-9563 ISSN 2363-9571 (electronic)
Current Topics in Neurotoxicity
ISBN 978-3-030-08769-2 ISBN 978-3-319-78707-7 (eBook)
https://doi.org/10.1007/978-3-319-78707-7

Softcover re-print of the Hardcover 1st edition 2018

Printed on acid-free paper

This Springer imprint is published by the registered company Springer International Publishing AG part of Springer Nature
The registered company address is: Gewerbestrasse 11, 6330 Cham, Switzerland

Preface

Over the last decade, a rapidly increasing number of novel psychoactive substance (NPS), often marketed as "designer drugs", "legal highs", "herbal highs", "research or intermediate chemicals", and "laboratory reagents", has appeared on the drug market worldwide in an effort to bypass-controlled substance legislation. Products containing those compounds can be purchased online via website distributors, from retail and not-retail vendors. The escalating popularity and use of NPSs has already resulted in a global drug abuse crisis.

NPS represent a heterogenous family of substances. One of the most prevalent NPS's groups is designer cathinones, endowed with psychostimulatory action, and abused for effects similar to amphetamine, cocaine, methamphetamine, and methylenedioxymethamphetamine. Compounds from this class were first detected in Europe in 2004, and since that time more than 130 new cathinones have been identified and reported to the European Monitoring Centre for Drugs and Drug Addiction. The rapid and extensive worldwide rise of synthetic cathinone abuse is attracting increasing attention, due to many intoxications and overdose deaths.

In light of the recent emergence of designer cathinones as substances of abuse, this book provides readers with the background of their availability, legal status, prevalence, motivation, and pattern of use. The book also presents a timely and comprehensive overview of the known mechanisms of action and pharmacokinetic properties of cathinones, analytical methods applied for the detection of parent compounds, and their metabolites in biological samples. Special emphasis is given to acute and chronic toxicity associated with the use of these substances.

I believe that this multiauthored book would be a valuable reference not only for scientists, clinicians, and academics but also for readers representing a different professional background who are involved in educational–prophylactic activities directed to harm reduction of psychoactive compounds.

I am truly indebted to all of the contributors of this book. I would like to thank Prof. Richard M. Kostrzewa, Editor-in-Chief of the Springer Neuroscience/Neurotoxicity Book Series, for entrusting me with the honorable function as this volume's editor.

Łódź, Poland Jolanta B. Zawilska
2018

Contents

Contributors

Piotr Adamowicz Institute of Forensic Research, Kraków, Poland

Trevor Archer Department of Psychology, University of Gothenburg, Gothenburg, Sweden

Amine Benyamina Addiction Research and Treatment Center, Paul Brousse Hospital (AP-HP), Paris-Sud University, Villejuif, France; CESP, Paris, France

John M. Corkery Psychopharmacology, Drug Misuse and Novel Psychoactive Substances Research Unit, School of Life and Medical Sciences, University of Hertfordshire, Hatfield, UK

Taylor A. Gentile Lewis Katz School of Medicine, Temple University, Philadelphia, PA, USA

Krystyna Gołembiowska Institute of Pharmacology, Polish Academy of Sciences, Kraków, Poland

Amira Guirguis Psychopharmacology, Drug Misuse and Novel Psychoactive Substances Research Unit, School of Life and Medical Sciences, University of Hertfordshire, Hatfield, UK

Katarzyna Kamińska Institute of Pharmacology, Polish Academy of Sciences, Kraków, Poland

Laurent Karila Addiction Research and Treatment Center, Paul Brousse Hospital, Paris-Sud University, INSERM U1000, Villejuif, France

Erin Kim Lewis Katz School of Medicine, Temple University, Philadelphia, PA, USA

Richard M. Kostrzewa Department of Biomedical Sciences, Quillen College of Medicine, East Tennessee State University, Johnson City, TN, USA

Ali Murad Lewis Katz School of Medicine, Temple University, Philadelphia, PA, USA

John W. Muschamp Lewis Katz School of Medicine, Temple University, Philadelphia, PA, USA

Laura Orsolini Psychopharmacology, Drug Misuse and Novel Psychoactive Substances Research Unit, School of Life and Medical Sciences, University of Hertfordshire, Hatfield, UK

Duccio G. Papanti Psychopharmacology, Drug Misuse and Novel Psychoactive Substances Research Unit, School of Life and Medical Sciences, University of Hertfordshire, Hatfield, UK

Nilesh B. Patel Department of Medical Physiology, School of Medicine, College of Health Sciences, University of Nairobi, Nairobi, Kenya

Scott M. Rawls Lewis Katz School of Medicine, Temple University, Philadelphia, PA, USA

Fabrizio Schifano Psychopharmacology, Drug Misuse and Novel Psychoactive Substances Research Unit, School of Life and Medical Sciences, University of Hertfordshire, Hatfield, UK

Linda D. Simmler Department of Basic Neurosciences, University of Geneva, Geneva, Switzerland

Steven J. Simmons Lewis Katz School of Medicine, Temple University, Philadelphia, PA, USA

Jakub Wojcieszak Department of Pharmacodynamics, Medical University of Łódź, Łódź, Poland

Kei Zaitsu Department of Legal Medicine and Bioethics, Nagoya University Graduate School of Medicine, Showa-ku, Nagoya, Japan

Jolanta B. Zawilska Department of Pharmacodynamics, Medical University of Łódź, Łódź, Poland

Dariusz Zuba Institute of Forensic Research, Kraków, Poland

Chapter 1
Synthetic Cathinones: Neurotoxic Health Hazards and Potential for Abuse

Trevor Archer and Richard M. Kostrzewa

Abstract Synthetic cathinones (SCs) represent an alternative to the prevailing array of drugs of abuse, with this being attributable to a variety of factors pertaining to reward/reinforcement potency, apparent but not necessarily lower toxicity risk, and easy access in "alternative" marketplaces. SCs retain distinctive neuropharmacological profiles and the induction of a disparate behavioral spectrum relating to locomotor activity, learning effectiveness, anxiety, thermoregulation, and abuse liability.

Keywords Synthetic cathinones · Reward · Self-administration Neurotoxicity · Health hazards

Cathinone derivatives, or synthetic cathinones (SCs), are sold as "bath salts" or as a "jewelry cleaner" (Drug Enforcement Administration, Department of Justice 2014a, b) in order to skirt law enforcement regulations on the sale of substances of abuse. The SCs, in reality, are β-keto analogs of commonly abused amphetamines (AMPHs), displaying pharmacological effects analogous to those of cocaine and AMPHs. The SCs are among the most common and prevalent groups of new psychoactive substances (NPSs), accommodating more than 620 different compounds (Guirguis et al. 2017; Hayashi et al. 2017; Romanek et al. 2017) that produce unique neuropharmacological effects, and are abused by a wide range of both infected (e.g., HIV, human immunodeficiency virus) and noninfected individuals, accounting for more than 30% of the criminal offense NPS drug seizures in European countries, despite lack of evidence for "epidemic proportions" in the United States several years ago (Stogner and Miller 2013).

T. Archer (✉)
Department of Psychology, University of Gothenburg, PO Box 500
430 50 Gothenburg, Sweden
e-mail: trevor.archer@psy.gu.se

R. M. Kostrzewa
Department of Biomedical Sciences, Quillen College of Medicine,
East Tennessee State University, Johnson City, TN 37614, USA

J. B. Zawilska (ed.), *Synthetic Cathinones*, Current Topics in Neurotoxicity 12,
https://doi.org/10.1007/978-3-319-78707-7_1

Nonetheless, strategies for surveillance, detection, and assessment continue development apace (Alvarez et al. 2017), since the uniquely reinforcing subjective effects linked to NPS and SCs' usage induce harmful effects that may be severe and/or lethal (Assi et al. 2017). The extraction efficiency and stability parameters of these compounds have been reasonably documented (Miller et al. 2017). "Designer drugs" (e.g., SCs and synthetic cannabinoids) continue to arise and proliferate at alarming levels, culminating in greater numbers of lethal overdose cases (Pourmand et al. 2018). Although an abundance of evidence has been presented to describe the characteristics of SCs as psychostimulants, their behavioral effects on locomotion, memory, and potential for abuse, the expressions of their neurotoxic properties await comprehensive elucidation (Angoa-Pérez et al. 2017). SCs may be counted also among the performance and image enhancing drugs which include anabolic androgenic steroids, AMPH-like substances, coffee and caffeine derivatives, and ephedrine (Piacentino et al. 2017) whose users are oft associated with higher levels of physical exercise, higher daily intake of coffee and psychotropic medication (e.g., benzodiazepines), more SCID (i.e., semi-structured interviews), diagnoses of psychiatric disorders, especially substance use disorder, body dysmorphic disorder, eating disorders, generalized anxiety disorder, greater body image concern inventory scores, potentially higher risk for body dysmorphic disorder, and higher Sick, Control, One, Fat, Food Eating Disorder Screening Test scores, which implies higher risk levels for the body image disorders and eating disorders. In a study of individual demographics, drug use, mental health characteristics, and criminal histories of NPS users as opposed to non-NPS users encountered in an offender population that was entering a drug treatment program (N = 8791), it was observed that the SCs users offended more often were more likely to have experienced homelessness, and to have lived in a metropolitan area than the latter (Smith et al. 2017), displaying also greater past-year drug use, including substances not readily detected by standard urine analysis (e.g., hallucinogens, alcohol, and inhalants). Individuals' greater anxiety symptom measures and those who showed the drinking-to-intoxication habit displayed also an increased likelihood of NPS usage, whereas older individuals or those initiating drug use at higher age levels were less likely to report NPS usage. These analyses imply that the use of NPS may offer a marker of more severe use patterns in an offender sample.

The SCs have been placed into three neuropharmacological categories on the basis of their influence upon monoamine transporters (Simmler et al. 2014), with similarities to the manner in which unlawful psychostimulants are categorized: (i) reuptake blocking agents of monoamines with "cocaine-like" effects, some of which exert greater dopaminergic than serotoninergic properties (Simmler and Liechti 2017); (ii) releasing agents with a "methamphetamine (METH)-like" action with additional reuptake blocking properties and higher catecholamine selectivity; (iii) empathogenic agents, i.e., inducing experiences of emotional communion, oneness, relatedness, emotional openness relating to empathy and sympathy, with an "MDMA-like" (MDMA, 3,4–methylenedioxymethamphetamine) action with both reuptake blocking and releasing actions, although the serotonergic releasing property exceeds that of the catecholaminergic. The reinforcing strength of these compounds may be formidable: for example, 3,4-methylenedioxypyrovalerone (MDPV), a catecholamine reuptake inhibitor, exerts a reinforcing capacity that exceeds that of METH in drug self-administering rats (Aarde et al. 2013; Aarde and Taffe 2017; Nguyen et al. 2016), with also powerful rewarding and reinforcing effects relative to cocaine (Simmons et al. 2016; Watterson and Olive 2017). Furthermore, the emerging SCs, α-pyrrolidinopentiophenone (α-PVP), 4-methyl-*N*-ethylcathinone, and 4-methyl-α-pyrrolidinopropiophenone are dopamine reuptake blockers lacking monoamine release or serotonin reuptake blockade. They were observed to be more effective reinforcers than METH, thereby implying a high-risk abuse potential (Huskinson et al. 2017). Additionally, the two SCs, α-PVP and 4-methyl-α-pyrrolidinopropiophenone, induced subjective effects in drug discrimination learning to METH (Naylor et al. 2015). In the context of behavioral and molecular effects, MDPV (higher potency for dopamine transporter (DAT) blockade) displayed more profound psychomotor effects than α-PVP, although both affected spontaneous- and stimulated-motor behavior (Giannotti et al. 2017). Each obtained also a differential modulation of immediate early genes in the frontal lobe, striatum, and hippocampus in laboratory mice, which partially explains the differential levels of toxicity.

Methedrone, methylone, and MDPV appear to be most prevalent and toxicologically problematic SCs. Each blocks the DAT, SERT (serotonin transporter) and NET (norepinephrine transporter)—thereby producing cocaine-like actions. Simmler's chapter on "Monoamine transporter and receptor interaction profiles of synthetic cathinones", outlines a comprehensive review of the pharmacological profiles of these respective SCs. Similarities in chemical structure between the SCs, AMPH, cocaine, and MDMA have produced like-mechanisms of action (Katz et al. 2014). The chapter by Patel, "Khat—a natural precursor of synthetic cathinones", outlines the uses and abuses of khat among inhabitants of Southwestern and East Africa, and also khat distribution and cultivation and biosynthetic pathway involving cathinone as a precursor molecule. This account is notable in view of khat chewing's marked association with shorter working hours; higher average number of significant road traffic accidents; and higher frequency of traffic violations among professional drivers in Saudi Arabia (Awadalla and Suwaydi 2017). Khat, although possessing an antidepressant effect (Alfaifi et al. 2017), has strong associations with tobacco, alcohol, and illicit drug abuse (Nakajima et al. 2017), and is a strong predictor of poor academic performance among university students (Mekonen et al. 2017). The chapter by Zuba and Adamowicz, "Analytical methods used for identification and determination of synthetic cathinones and their metabolites", describes several techniques to analyzed the different drugs seized by civil authorities, including gas chromatography—mass spectrometry, Fourier-transformed infrared spectrometry, nuclear magnetic resonance, liquid chromatography (with tandem mass spectrometers); and using blood serum, dried blood spots, urine, hair, oral fluid and postmortem body tissues and following a myriad of tissue preparations analysis through liquid chromatography—high-resolution spectrometry. Among the seized drugs they have identified, α-PVP which subsisted in 84% of the samples (Byrska et al. 2017) in addition to the large quantity of other dangerous NPSs (Styszko et al. 2016; see also Adamowicz et al. 2014).

There is an increasing need to comprehend the neuropharmacological components underpinning the reinforcing properties and neurotoxicological status of the SCs. Thus, the biological influences of mephedrone (4-methyl-*N*-methylcathinone)—as derived from drug user narrative experiences, epidemiological data, clinical case studies, toxicological findings, and animal laboratory studies—gives insight into human pharmacokinetic and pharmacodynamic properties and toxicological profile (Papaseit et al. 2017). MDPV, a potent uptake inhibitor at the DAT and NET, is rapidly metabolized to 3,4-dihydroxypyrovalerone and 4-hydroxy-3-methoxypyrovalerone in vivo, with notable drug-induced motor activation that is correlated positively with the plasma concentrations of parent MDPV (Baumann et al. 2017). In this regard, the ongoing emergence of structural-activity relationships involving the (*a*) characterization of the terminal amine, (*b*) size of the α-substituent, stereochemistry, and (*c*) presence and position of aromatic substituents establish the impingement of action sequences (such as the release agents or reuptake inhibitors) and transporter selectivity (Glennon and Dukat 2017). The chapter by Gołembiowska and Kamińska, "Effects of synthetic cathinones on brain neurotransmitters", is a review of SCs' profound influences upon brain levels of neurotransmitters, as expressed by the potent and dose-dependent elevations of striatal and accumbens DA and 5-HT. Mephedrone similarly elevated DA levels in cultures of striatum and nucleus accumbens, and reduced 5-HT levels in hippocampus and frontal cortices. Grecco et al. (2017) studied the in vivo blood plasma and CNS pharmacokinetics of three SCs with differing length of their α-alkyl chain: methylone ($-CH_3$), butylone ($-CH_2CH_3$), and pentylone ($-CH_2CH_2CH_3$), exhibiting a range of effects. Analysis of urine samples of MDPV users, 87% male (age range 23–47 years), revealed the presence of MDPV and its metabolites 2′-oxo-MDPV, demethylenyl-MDPV, demethylenyl-methyl-MDPV, demethylenyl-oxo-MDPV, demethylenyl-methyl-oxo-MDPV, and demethylenyl-methyl-*N*,*N*-bisdealkyl-MDPV (Grapp et al. 2017). The chapter by Zaitsu, "Metabolism of synthetic cathinones", outlines the stages of phase I and phase II metabolic pathways which were identified not only for the presence of hydrolyzed-and/or hydroxylated metabolites but also for non-changed cathinone derivatives, e.g., as observed in the analysis of urinary phase I metabolites of α-pyrrolidinobutiophenone (α-PBP) in humans (Matsuta et al. 2015).

The SCs, MDPV and α-PVP, induced a neuropharmacological action related to block of synaptic DAT and NET (Stanciu et al. 2017). These compounds have a short duration of action which results in the situation that drug users readminister the SC both frequently and compulsively, and by which they induce and persist in a behavior resembling delirium. These subjects display disorganization, delusional thinking, and hallucinations which eventually unfolds into severe "consequence-excited delirium", often characterized by intensive paranoid behavior leading to extremely violent outbursts that must be treated with the use of pronounced and forceful physical preventative interventions, with restraints linked to increased mortality. In the chapter by Karila and Benyamina, "The effects and risks associated with synthetic cathinones use in humans", the individual- and dose-dependency effects of SCs are discussed clinically with regard to user expressions of euphoria, empathy, increased sexual performance, and enlightened sociability but also with bouts of panic attacks. Similarly, SC abuse is associated with serious physiological and psychological risk factors, including negative physical symptoms (e.g., tachycardia), neuropsychiatric symptoms (e.g., mood disturbances, and neurological symptoms (e.g., hyperactivity, adverse effects that may/may not culminate in violence, homicidal combative behavior, self-mutilatory behavior, coma, and death. As described by Cockery et al. in the chapter, "Synthetic cathinones—prevalence and motivations for use", consumption of these SCs appears to be related to the imbalance/dysregulation occurring over a labyrinthine variety of pre- and post-synaptic neurotransmitter systems and receptors, with particular focus on monoaminergic pathways, but also cannabinoid pathways. These considerations require the careful monitoring of certain behaviors, e.g., self/outward-directed aggression and agitation, in order to obtain insights into users' drug habits and thus achieve an acute management of NPS adverse events (Schifano et al. 2016, 2017).

The chapter by Simmons et al. "Behavioral profiles and underlying transmitters/circuits of cathinone-derived psychostimulant drugs of abuse" is a deep summary of the behavioral effects following synthetic cathinone in animals. It also helps us to understand the transmitters and circuits that underlie these effects. A low dose of 4-methylethcathinone (10 mg/kg) was used to induce conditioned place preference and reinstatement, while a higher dose (30 mg/kg) increased locomotor activity (Xu et al. 2016). Chronic 4-methylethcathinone treatment induced delayed and attenuated sensitization in comparison with METH, with the former increasing anxiety and the latter decreasing anxiety in the elevated plus maze. When administering the second-generation SCs, α-PVP, α-pyrrolidinopropiophenone, 3,4-methylenedioxy-α-pyrrolidinobutiophenone, and 3,4-methylenedioxy-α-pyrrolidinopropiophenone, to laboratory rats using fixed-ratio schedules of reinforcement to maintain operant behavior performance, it was noted that the observed rank order potency to maintain responding during a fixed-ratio 5 schedule of reinforcement (i.e., the rewarding effect of each compound) was dependent on potency of DAT inhibition (Gannon et al. 2017a). Through the derivation of full dose–response curves under three different schedules of reinforcement for self-administration, it was shown that MDPV was approximately tenfold more potent and approximately threefold more effective, despite greater variability, at maintaining responding, versus cocaine (Gannon et al. 2017b). MDPV exerts a reinforcing influence that, seemingly, is markedly more effective than either cocaine or METH; the drug also possesses a remarkable capacity to construct an enduring phenotype in the rat model applied, which is expressed by the surprisingly high levels of drug intake. Uniquely, the existence of both a potent rewarding influence and an aversive property has been assigned to the SCs (King and Riley 2017).

The chapter by Zawilska and Wojcieszak, "Novel psychoactive substances: classification and general information", presents and evaluates the upsurge and proliferation, origin, and classification of the SCs, as well as their "attractiveness" to users/abusers. The review outlines the neuropharmacological properties since these may build a framework for the essential neurotoxicological aspects of these compounds. Nevertheless, several lines of evidence have asserted that, taken together, the commonly used SCs were experienced to be less rewarding than ecstasy and cocaine in terms of pleasurable effects and prognosis for future use (Matthews et al. 2017), although the negative effects of these drugs, whether "come-down" or acute, were rated at a lower level of intensity. The SCs exert an assortment of adverse effects that include various forms of psychosis, sometimes severe; neurologic and movement dysfunction, cognitive deficits, cardiovascular and respiratory complications (e.g., Barrios et al. 2016), and a diversity of other health problems such as drug dependence, vomiting and nausea, acid–base imbalance, electrolytic imbalance and kidney injury, rhabdomyolysis, vomiting and muscle pain, liver failure, hepatitis C, and diaphoresis (Dolengevich-Segal et al. 2016), as well as unpredictable behavioral changes to agitation, aggression, and violence (Gray et al. 2016). The authors have analyzed the cytotoxic potential of both first- and second-generation SCs by assessing effects from the two new generation pyrovalerones, α-pyrrolidinovalerothiophenone (α-PVT), and α-pyrrolidinooctanophenone (PV9), which represent the most potent cytotoxic compounds—producing pronounced mitochondrial dysfunction; while α-pyrrolidinooctanophenone produces marked damage to cell membranes (Wojcieszak et al. 2016). Not least among the adverse effects of the SCs is the associated marked hyperthermia (Lantz et al. 2017; Valente et al. 2016), a dose-dependent effect attenuated by the α- and β-adrenoceptor antagonist, carvedilol (Zona et al. 2016). In a German web-based survey of volunteer users of "bath salts" (N = 99, 74% male participants), SC-induced confusion was the most commonly occurring complication, followed by violent thoughts either acted upon others or themselves, high-risk sexual behavior, problems with law enforcement for selling cathinones, criminal behavior; and finally, most of the SCs users were users of other drugs including alcohol (Schmoll et al. 2017), and generally were motivated by opportunistic reasons, e.g., availability or assumed legality (Sutherland et al. 2017). The use of methylone (most prevalent of SCs in this context) and other drugs in electronic dance music of nightclubs and festivals in New York City during 2015, using a variation of time-space sampling and a hair-testing procedure, was studied and implied that attendees were a high-risk group for use, prevalence and necessity for harm-reduction services (Palamar et al. 2016, 2017a, b, and c; Salomone et al. 2017).

In conclusion, the unfolding evidence on structure and function of primary, secondary and, eventually, the tertiary generations of SCs will continue to depict these compounds as highly dangerous, and from a legislative viewpoint, rather sordid drugs of abuse with manifest levels of neurotoxicity to general and cerebral aspects of health.

References

Aarde SM, Huang PK, Creehan KM et al (2013) The novel recreational drug 3,4-methylenedioxypyrovalerone (MDPV) is a potent psychomotor stimulant: self-administration and locomotor activity in rats. Neuropharmacology 71:130–140. https://doi.org/10.1016/j.neuropharm.2013.04.003

Aarde SM, Taffe MA (2017) Predicting the abuse liability of entactogen-class, new and emerging psychoactive substances via preclinical models of drug self-administration. Curr Top Behav Neurosci 32:145–164. https://doi.org/10.1007/7854_2016_54

Adamowicz P, Zuba D, Byrska B (2014) Fatal intoxication with 3-methyl-N-methylcathinone (3-MMC) and 5-(2-aminopropyl)benzofuran (5-APB). Forensic Sci Int 245:126–132. https://doi.org/10.1016/j.forsciint.2014.10.016

Alfaifi H, Abdelwahab SI, Mohan S et al (2017) *Catha edulis* Forsk. (Khat): evaluation of its antidepressant-like activity. Pharmacogn Mag 13(Suppl 2):S354–S358. https://doi.org/10.4103/pm.pm_442_16

Alvarez JC, Fabresse N, Larabi IA (2017) Prevalence and surveillance of synthetic cathinones use by hair analysis: an update review. Curr Pharm Des. https://doi.org/10.2174/1381612823666170704124156

Angoa-Pérez M, Anneken JH, Kuhn DM (2017) Neurotoxicology of synthetic cathinone analogs. Curr Top Behav Neurosci 32:209–230. https://doi.org/10.1007/7854_2016_21

Assi S, Gulyamova N, Ibrahim K et al (2017) Profile, effects, and toxicity of novel psychoactive substances: a systematic review of quantitative studies. Hum Psychopharmacol 32(3). https://doi.org/10.1002/hup.2607

Awadalla NJ, Suwaydi HA (2017) Prevalence, determinants and impacts of khat chewing among professional drivers in Southwestern Saudi Arabia. East Mediterr Health J 23(3):189–197

Barrios L, Grison-Hernando H, Boels D et al (2016) Death following ingestion of methylone. Int J Legal Med 130(2):381–385. https://doi.org/10.1007/s00414-015-1212-4

Baumann MH, Bukhari MO, Lehner KR et al (2017) Neuropharmacology of 3,4-methylenedioxypyrovalerone (MDPV), its metabolites, and related analogs. Curr Top Behav Neurosci 32:93–117. https://doi.org/10.1007/7854_2016_53

Byrska B, Stanaszek R, Zuba D (2017) Alpha-PVP as an active component of herbal highs in Poland between 2013 and 2015. Drug Test Anal 9(8):1267–1274. https://doi.org/10.1002/dta.2151

Dolengevich-Segal H, Rodríguez-Salgado B, Gómez-Arnau J, Sánchez-Mateos D (2016) Severe psychosis, drug dependence, and hepatitis C related to slamming mephedrone. Case Rep Psychiatry 2016:8379562. https://doi.org/10.1155/2016/8379562

Drug Enforcement Administration, Department of Justice (2014a) Schedules of controlled substances: temporary placement of 10 synthetic cathinones into Schedule I, Final Order. Fed Regist 79:12938–12943

Drug Enforcement Administration, Department of Justice (2014b) Schedules of controlled substances: extension of temporary placement of 10 synthetic cathinones into Schedule I of the Controlled Substances Act. Final order. Fed Regist 81:11429–11431

Gannon BM, Galindo KI, Mesmin MP et al (2017a) Relative reinforcing effects of second-generation synthetic cathinones: acquisition of self-administration and fixed ratio dose-response curves in rats. Neuropharmacology. pii: S0028-3908(17)30385-4. https://doi.org/10.1016/j.neuropharm.2017.08.018

Gannon BM, Galindo KI, Rice KC, Collins GT (2017b) Individual differences in the relative reinforcing effects of 3,4-methylenedioxypyrovalerone under fixed and progressive ratio schedules of reinforcement in rats. J Pharmacol Exp Ther 361(1):181–189. https://doi.org/10.1124/jpet.116.239376

Giannotti G, Canazza I, Caffino L et al (2017) The cathinones MDPV and α-PVP elicit different behavioral and molecular effects following acute exposure. Neurotox Res. https://doi.org/10.1007/s12640-017-9769-y

Glennon RA, Dukat M (2017) Structure-activity relationships of synthetic cathinones. Curr Top Behav Neurosci 32:19–47. https://doi.org/10.1007/7854_2016_41

Grapp M, Kaufmann C, Ebbecke M (2017) Toxicological investigation of forensic cases related to the designer drug 3,4-methylenedioxypyrovalerone (MDPV): detection, quantification and studies on human metabolism by GC-MS. Forensic Sci Int 273:1–9. https://doi.org/10.1016/j.forsciint.2017.01.021

Gray R, Bressington D, Hughes E, Ivanecka A (2016) A systematic review of the effects of novel psychoactive substances 'legal highs' on people with severe mental illness. J Psychiatr Ment Health Nurs 23(5):267–281. https://doi.org/10.1111/jpm.12297

Grecco GG, Kisor DF, Magura JS, Sprague JE (2017) Impact of common clandestine structural modifications on synthetic cathinone "bath salt" pharmacokinetics. Toxicol Appl Pharmacol 328:18–24. https://doi.org/10.1016/j.taap.2017.05.010

Guirguis A, Corkery JM, Stair JL et al (2017) Intended and unintended use of cathinone mixtures. Hum Psychopharmacol 32(3). https://doi.org/10.1002/hup.2598

Hayashi K, Wakabayashi C, Ikushima Y, Tarui M (2017) High prevalence of quasi-legal psychoactive substance use among male patients in HIV care in Japan: a cross-sectional study. Subst Abuse Treat Prev Policy 12(1):11. https://doi.org/10.1186/s13011-017-0097-2

Huskinson SL, Naylor JE, Townsend EA et al (2017) Self-administration and behavioral economics of second-generation synthetic cathinones in male rats. Psychopharmacology 234 (4):589–598. https://doi.org/10.1007/s00213-016-4492-6

Katz DP, Bhattacharya D, Bhattacharya S et al (2014) Synthetic cathinones: "a khat and mouse game". Toxicol Lett 229(2):349–356. https://doi.org/10.1016/j.toxlet.2014.06.02

King HE, Riley AL (2017) The affective properties of synthetic cathinones: role of reward and aversion in their abuse. Curr Top Behav Neurosci 32:165–181. https://doi.org/10.1007/7854_2016_32

Lantz SM, Rosas-Hernandez H, Cuevas E et al (2017) Monoaminergic toxicity induced by cathinone phthalimide: an in vitro study. Neurosci Lett 655:76–81. https://doi.org/10.1016/j.neulet.2017.06.059

Matthews A, Sutherland R, Peacock A et al (2017) I like the old stuff better than the new stuff? Subjective experiences of new psychoactive substances. Int J Drug Policy 40:44–49. https://doi.org/10.1016/j.drugpo.2016.11.004

Matsuta S, Shima N, Kamata H et al (2015) Metabolism of the designer drug α-pyrrolidinobutiophenone (α-PBP) in humans: identification and quantification of the phase I metabolites in urine. Forensic Sci Int 249:181–188. https://doi.org/10.1016/j.forsciint.2015.02.004

Mekonen T, Fekadu W, Mekonnen TC, Workie SB (2017) Substance use as a strong predictor of poor academic achievement among university students. Psychiatry J. 2017:7517450. https://doi.org/10.1155/2017/7517450

Miller B, Kim J, Concheiro M (2017) Stability of synthetic cathinones in oral fluid samples. Forensic Sci Int 274:13–21. https://doi.org/10.1016/j.forsciint.2016.11.034

Nakajima M, Molla K, Belachew B et al (2017) Khat use is associated with tobacco, alcohol, and illicit drug use: a cross-sectional examination in the United States. J Psychoactive Drugs 49 (5):413–419. https://doi.org/10.1080/02791072.2017.1342155

Naylor JE, Freeman KB, Blough BE et al (2015) Discriminative-stimulus effects of second generation synthetic cathinones in methamphetamine-trained rats. Drug Alcohol Depend 149:280–284. https://doi.org/10.1016/j.drugalcdep.2015.02.002

Nguyen JD, Aarde SM, Cole M et al (2016) Locomotor stimulant and rewarding effects of inhaling methamphetamine, MDPV, and mephedrone via electronic cigarette-type technology. Neuropsychopharmacology 41(11):2759–2771. https://doi.org/10.1038/npp.2016.88

Palamar JJ, Acosta P, Sherman S et al (2016) Self-reported use of novel psychoactive substances among attendees of electronic dance music venues. Am J Drug Alcohol Abuse 42(6):624–632. https://doi.org/10.1080/00952990.2016.1181179

Palamar JJ, Acosta P, Calderón FF et al (2017a) Assessing self-reported use of new psychoactive substances: the impact of gate questions. Am J Drug Alcohol Abuse 43(5):609–617. https://doi.org/10.1080/00952990.2017.1322094

Palamar JJ, Acosta P, Cleland CM (2017b) Attitudes and beliefs about new psychoactive substance use among electronic dance music party attendees. Subst Use Misuse 1–10. https://doi.org/10.1080/10826084.2017.1327980

Palamar JJ, Salomone A, Gerace E et al (2017c) Hair testing to assess both known and unknown use of drugs amongst ecstasy users in the electronic dance music scene. Int J Drug Policy 48:91–98. https://doi.org/10.1016/j.drugpo.2017.07.010

Papaseit E, Moltó J, Muga R et al (2017) Clinical pharmacology of the synthetic cathinone mephedrone. Curr Top Behav Neurosci 32:313–331. https://doi.org/10.1007/7854_2016_61

Piacentino D, Kotzalidis GD, Longo L et al (2017) Body image and eating disorders are common among professional and amateur athletes using performance and image enhancing drugs: a cross-sectional study. J Psychoactive Drugs 49(5):373–384. https://doi.org/10.1080/02791072.2017.1359708

Pourmand A, Mazer-Amirshahi M, Chistov S et al (2018) Designer drugs: review and implications for emergency management. Hum Exp Toxicol 37(1):94–101. https://doi.org/10.1177/0960327117698542

Romanek K, Stenzel J, Schmoll S et al (2017) Synthetic cathinones in Southern Germany—characteristics of users, substance-patterns, co-ingestions, and complications. Clin Toxicol (Phila) 55(6):573–578. https://doi.org/10.1080/15563650.2017.1301463

Salomone A, Palamar JJ, Gerace E et al (2017) Hair testing for drugs of abuse and new psychoactive substances in a high-risk population. J Anal Toxicol 41(5):376–381. https://doi.org/10.1093/jat/bkx020

Schifano F, Orsolini L, Papanti D, Corkery J (2017) NPS: medical consequences associated with their intake. Curr Top Behav Neurosci 32:351–380. https://doi.org/10.1007/7854_2016_15

Schifano F, Papanti GD, Orsolini L, Corkery JM (2016) Novel psychoactive substances: the pharmacology of stimulants and hallucinogens. Expert Rev Clin Pharmacol 9(7):943–954. https://doi.org/10.1586/17512433.2016.1167597

Schmoll S, Romanek K, Stich R et al (2017) An internet-based survey of 96 German-speaking users of "bath salts": frequent complications, risky sexual behavior, violence, and delinquency. Clin Toxicol (Phila). https://doi.org/10.1080/15563650.2017.1353094

Simmler LD, Liechti ME (2017) Interactions of cathinone NPS with human transporters and receptors in transfected cells. Curr Top Behav Neurosci 32:49–72. https://doi.org/10.1007/7854_2016_20

Simmler LD, Rickli A, Hoener MC, Liechti ME (2014) Monoamine transporter and receptor interaction profiles of a new series of designer cathinones. Neuropharmacology 152–160. https://doi.org/10.1016/j.neuropharm.2013.11.008

Simmons SJ, Gregg RA, Tran FH et al (2016) Comparing rewarding and reinforcing properties between 'bath salt' 3,4-methylenedioxypyrovalerone (MDPV) and cocaine using ultrasonic vocalizations in rats. Addict Biol. https://doi.org/10.1111/adb.12479

Smith KE, Bunting AM, Staton M et al (2017) Examination of synthetic cannabinoid and cathinone use among a drug-using offender sample, 2013–2015. J Psychoactive Drugs 49 (5):436–445. https://doi.org/10.1080/02791072.2017.1361560

Stanciu CN, Penders TM, Gnanasegaram SA et al (2017) The behavioral profile of methylenedioxypyrovalerone (MDPV) and α-pyrrolidinopentiophenone (PVP)—a systematic review. Curr Drug Abuse Rev. https://doi.org/10.2174/1874473710666170321122226

Stogner JM, Miller BL (2013) Investigating the 'bath salt' panic: the rarity of synthetic cathinone use among students in the United States. Drug Alcohol Rev 32(5):545–549. https://doi.org/10.1111/dar.12055

Styszko K, Dudarska A, Zuba D (2016) The presence of stimulant drugs in wastewater from Krakow (Poland): a snapshot. Bull Environ Contam Toxicol 97(3):310–315. https://doi.org/10.1007/s00128-016-1869-5

Sutherland R, Bruno R, Peacock A et al (2017) Motivations for new psychoactive substance use among regular psychostimulant users in Australia. Int J Drug Policy 43:23–32. https://doi.org/10.1016/j.drugpo.2016.12.021

Valente MJ, Araújo AM, Silva R et al (2016) 3,4-Methylenedioxypyrovalerone (MDPV): in vitro mechanisms of hepatotoxicity under normothermic and hyperthermic conditions. Arch Toxicol 90(8):1959–1973. https://doi.org/10.1007/s00204-015-1653-z

Watterson LR, Olive MF (2017) Reinforcing effects of cathinone NPS in the intravenous drug self-administration paradigm. Curr Top Behav Neurosci 32:133–143. https://doi.org/10.1007/7854_2016_33

Wojcieszak J, Andrzejczak D, Woldan-Tambor A, Zawilska JB (2016) Cytotoxic activity of pyrovalerone derivatives, an emerging group of psychostimulant designer cathinones. Neurotox Res 30(2):239–250. https://doi.org/10.1007/s12640-016-9640-6

Xu P, Qiu Y, Zhang Y et al (2016) The effects of 4-methylethcathinone on conditioned place preference, locomotor sensitization, and anxiety-like behavior: a comparison with methamphetamine. Int J Neuropsychopharmacol 19(4). pii: pyv120. https://doi.org/10.1093/ijnp/pyv120

Zona LC, Grecco GG, Sprague JE (2016) Cooling down the bath salts: carvedilol attenuation of methylone and mephedrone mediated hyperthermia. Toxicol Lett 263:11–15. https://doi.org/10.1016/j.toxlet.2016.10.012

Chapter 2
Novel Psychoactive Substances: Classification and General Information

Jolanta B. Zawilska and Jakub Wojcieszak

Abstract Recent years have seen a steady growth in the availability and consumption of novel psychoactive substances (NPS) across the world. NPS are sold as legal, cheap and non-detectable substitutes for controlled drugs of abuse. The consumers are not normally provided with certain information on the composition of the product, or the recommended dosing and possible side effects. According to the EMCDDA, since 2009, tens of new compounds belonging to six major groups of NPS are introduced into the market each year. These major categories include psychostimulants, synthetic cannabimimetics, psychedelic compounds, dissociatives, synthetic opioids and benzodiazepine analogues. The number of compounds belonging to the two most popular groups, psychostimulatory cathinones and synthetic cannabimimetics, makes up to two-thirds of the total number of NPS. The great majority of NPS are synthetic compounds, and some of these were initially designed as potential medicines or pharmacological tools. NPS exert their actions by various pharmacological mechanisms such as the enhancement of central monoaminergic neurotransmission for psychostimulants, activation of cannabinoid receptors by synthetic cannabimimetics, activation of serotonin receptors by psychedelics, antagonism for NMDA receptors for dissociatives, and activation of μ-opioid and $GABA_A$ receptors by synthetic opioids and benzodiazepines, respectively. NPS can produce a wide spectrum of life-endangering side effects. Some of these actions are specific to a certain compound or group of compounds, while some can be produced by drugs from different classes of NPS. In general, NPS may impair cognitive functions, exacerbate or trigger mental disorders and cause severe neurological and cardiovascular symptoms. Significantly, there are specific antidotes for only two groups of NPS, opioids and benzodiazepines, and treatment of acute NPS intoxication is therefore usually symptomatic.

J. B. Zawilska (✉) · J. Wojcieszak
Department of Pharmacodynamics, Medical University of Łódź, Łódź, Poland
e-mail: jolanta.zawilska@umed.lodz.pl

J. Wojcieszak
e-mail: jakub.wojcieszak@umed.lodz.pl

J. B. Zawilska (ed.), *Synthetic Cathinones*, Current Topics in Neurotoxicity 12,
https://doi.org/10.1007/978-3-319-78707-7_2

Keywords Novel psychoactive substances · Psychostimulants · Synthetic cannabinomimetics · Hallucinogens · Benzodiazepines · Opioids

2.1 Introduction

Recent years have seen a rapid and continuous growth in the availability and use of novel psychoactive substances (NPS) across the world. An NPS is defined as 'a new narcotic or psychotropic drug, in pure form or in preparation, that is not controlled by the 1961 United Nations Single Convention on Narcotic Drugs or the 1971 United Nations Convention on Psychotropic Substances, but which may pose a public health threat comparable to that posed by substances listed in these conventions' (EMCDDA 2012; UNODC 2013a). These substances are psychoactive in that they stimulate or inhibit the function of the central nervous system (EMCDDA 2016a).

NPS include a wide range of products that can be purchased online via website distributors (both from freely accessible websites, and more recently, from the so-called dark/hidden web), from retail vendors ('head shops', 'smart shops' and 'smoke shops') and not-retail vendors (drug dealers) (EMCDDA 2015a, 2016b, c, 2017; UNODC 2013b, 2016, Papaseit et al. 2014; Vandrey et al. 2013; Wadsworth et al. 2017; Zawilska 2011). These substances are sold variously as 'legal/herbal highs', 'synthetic/designer drugs', 'smoking blends', 'bath salts', 'plant food', 'insect repellents', 'research chemicals', 'air fresheners', 'jewellery cleaner', 'pond water cleaner' and 'collectible products', with the disclaimer 'not for human consumption' or 'for research purposes only' to circumvent drug abuse legislation (EMCDDA 2015a; Zawilska 2011). Usually, the list of ingredients gives no indication of the presence of psychoactive compounds. Recommended dosing, side effects, warning and referrals to information sources are lacking. Importantly, product ingredients, both in terms of active compounds/substances present and their quantity, can markedly differ even within the same name brand from the same supplier. Forensic analytical reports indicate that NPS are often sold as combinations of different compounds. In addition to various NPS, the product may contain traditional drugs of abuse, such as amphetamine, 3,4-methylenedioxymethamphetamine (MDMA, ecstasy), cocaine and heroin, pharmaceutical products and other active adulterants, cutting agents or reaction by-products (Guirguis et al. 2017; Schifano et al. 2017; UNODC 2016; Zuba and Byrska 2013; Zuba et al. 2011). Altogether, consumers commonly do not know what active substance(s) they are using and in what dose(s).

2.2 The Number of NPS Is Rapidly Growing Around the World

Over the years, it has become well documented that new substances are produced very quickly in order to replace those that are placed under legal control by various countries. So not surprisingly, Helander and Bäckberg (2017) called NPS 'the Hydra monster of recreational drugs'. In 2013, the European Union (EU) drug market report emphasized that 'while the appearance of new drugs is not a new phenomenon, over the last few years there has been an unprecedented growth in their number, type and availability' (EMCDDA 2013) (see Fig. 2.1). Between 2008 and 2015, a total of 644 NPS were reported by 102 countries to the United Nations Office for Drugs and Crime, UNODC (UNODC 2016). The majority of countries and territories that reported the emergence of NPS up to December 2015 were from Europe, followed by Americas, Asia, Oceania and Africa (UNODC 2016). In line with these findings, by the end of December 2016, the European Union (EU) Early Warning System (EWS) had notified the appearance of 627 NPS (more than 70% of them had been made in the previous 5 years), and an almost eightfold increase in reported seizures of these drugs between 2008 and 2017 (EMCDDA 2017). In 2014 and 2015 only, around two new substances per week were reported to the EWS (EMCDDA 2016b).

The rapid growth of the NPS market has been facilitated by a number of factors, such as potent decrease in purity of ecstasy and cocaine, and the instability of the ecstasy market that occurred in the mid-2000s, improving technological capabilities in China and India, aggressive web-based marketing, increased trade via the Internet and well designed and targeted selling strategies (Brunt et al. 2011; EMCDDA 2016c, d; Huestis et al. 2017; Orsolini et al. 2015; UNODC 2016). Many NPS-containing products are sold in colourful packages professionally designed to attract and arouse interest. Numerous brand names remain consumers of illegal street drugs or their effects, e.g. 'Coco Power', 'Crack Inside', 'Snow Blow',

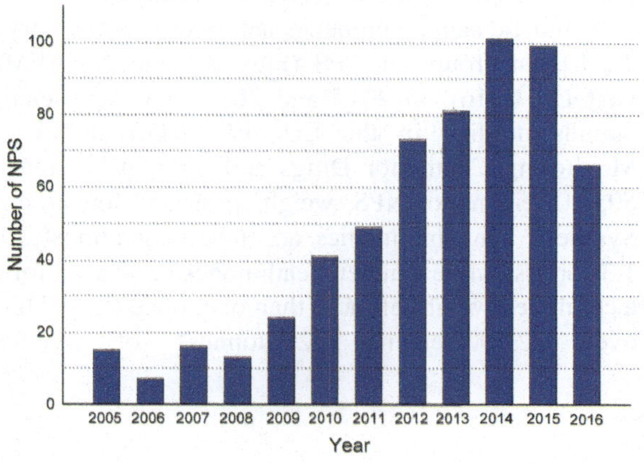

Fig. 2.1 Number of NPS notified to the European Monitoring Centre for Drugs and Drugs Addiction (EMCDDA 2016b, 2017)

'Xtacy', 'Rocket Fuel', 'Dynamite', 'Euphoria', 'Experience Chill', 'Herbal Dream', 'Synthetic Heroin', 'Fake morphine' and 'M-ket'. Other NPS brands evoke names of popular movies (e.g. 'Black mamba', 'Clockwork orange', 'Mad Alice' and 'Pink Panther'), cartoons (e.g. 'Scooby snax', 'He-Man' and 'Kryptonite'), animals (e.g. 'King Cobra', 'Snow Leopard' and 'Dragonfly') and geographical scenery (e.g. 'K2' and 'Moon rocks'). Pill formulations may be designed to appear as pharmaceuticals (Corazza et al. 2014).

2.3 Classification of NPS

NPS represent a heterogenous family of substances, including, among others, synthetic cannabimimetics, synthetic cathinones, phenethylamines, piperazines, ketamine- and phencyclidine-type substances, tryptamines, benzofurans, synthetic opioids (fentanyl analogues and compounds with a different chemical structure) and benzodiazepines (Abdulrahim and Bowden-Jones 2015; Papaseit et al. 2014; Schifano et al. 2017). Six major categories of NPS can be distinguished based on their psychopharmacological activity (Table 2.1):

(1) Psychostimulants that mimic the effects of amphetamine, cocaine, MDMA and methamphetamine;
(2) Synthetic cannabimimetics that mimic the effects of cannabis;
(3) Psychedelic compounds that exert effects similar to 'classical hallucinogens', such as LSD or psilocybin;
(4) Dissociatives that produce effects similar to ketamine or phencyclidine;
(5) Synthetic opioids that mimic the effects of morphine and heroin;
(6) Sedative/hypnotics benzodiazepine analogues.

While many of these substances often quickly disappear from the market, some, particularly synthetic cathinones such as mephedrone, have an established position as popular and attractive recreational drugs.

Synthetic cannabimimetics and psychostimulatory designer cathinones make up the largest groups of NPS (Figs. 2.2 and 2.3) (EMCDDA 2015b, 2016c, 2017; UNODC 2016); in 2015 and 2016 they represented almost two-thirds of compounds notified in the EU (EMCDDA 2017). According to the European Monitoring Centre for Drugs and Drug Addiction (EMCDDA), in 2014 almost 50,000 seizures of NPS, weighing around four tonnes, were made across Europe. Synthetic cannabimimetics accounted for almost 30,000 seizures and more than 1.3 tonnes, while synthetic cathinones accounted for more than 8000 seizures with a combined weight of more than one tonne (EMCDDA 2016b). The following year over 22,000 seizures (2.5 tonnes) of synthetic cannabimimetics and

Table 2.1 Main categories of novel psychoactive substances

Chemical group	Examples
Synthetic cannabinomimetics	
JWH compounds Naphtoylindoles Naphthylmethylindoles Naphthoylpyrroles Naphthylmethylindenes Phenacetylindoles	JWH-015, JWH-018, JWH-073, JWH-210 JWH-175 JWH-147 JWH-176 JWH-203, JWH-250, JWH-253
Amantoyindoles	AB-001
Aminoalkylindoles	WIN 55,212-2
Benzoylindoles	AM-694, AM-2201, AM-679, RCS-4
Cyclohexylphenols	CP 47,497 and its homologues
Classical cannabinoids (dibenzopyrans)	HU-210
Indazoles	AKB-48 (APINACA), ADB-PINACA, MDMB-CHMICA
TMCP compounds	TMCP-H, TMCP-018, TMCP-2201, TMCP-200, TMCP-1220
Other	UR-144, 5F-UR-144 (XLR-11), PB-22, BB-22 (QUCHIC), AB-FUBINACA
Psychostimulants	
Synthetic cathinones (β-keto-α-methyl-phenylalkylamines; β-keto amphetamines)	Butylone, ethcatinone, mephedrone, 4-methylethcathinone (4-MEC), 3-fluoromethcathinone (3-FMC), flephedrone, naphyrone, 3,4-methylenodioxypyrovalerone (MDPV), α-pyrovalerone (α-PVP)
Piperazines	1-Benzylpiperazine (BZP), 1-(3-trifluoromethylphenyl)piperazine (TFMPP), 1-(4-methoxyphenyl)piperazine (MeOPP), 1-(4-chlorophenyl)piperazine (mCPP), 1-(4-fluorophenyl)piperazine (pFPP)
Piperidines and pyrrolidines	2-(Diphenylmethyl) piperidine (desoxypipradrol, 2-DPMP), diphenyl-2-pyrrolidinylmethanol (diphenylprolinol; D2PM)
Tryptamines	5-(2-aminopropyl)indole (5-IT, 5-API)
2,5-Dimethoxyamphetamines	4-Bromo-2,5-dimethoxyamphetamine (DOB), 4-iodo-2,5-dimethoxyamphetamine (DOI), 2,5-dimethoxy-4-methylamphetamine (DOM), 2,4,5-trimethoxyamphetamine (TMA-2)
2-Aminoindanes	5-Iodo-2-aminoindane (5-IAI), 5,6-methylenodioxy-2-aminoindane (MDAI), 5,6-methylenodioxy-*N*-methyl-2-aminoindane (MMDAI)
Benzofurans	5-(2-Aminopropyl)benzofuran (5-APB), 6-(2-aminopropyl)benzofuran (6-APB), 1-benzofuran-5-yl)-*N*-ethylpropan-2-amine (5-EAPB)

(continued)

Table 2.1 (continued)

Chemical group	Examples
Hallucinogens **A. Psychedelics**	
Tryptamines	
1. *Simple tryptamines*	
1A. Without modification of the indole ring	α-Ethyltryptamine (AET), *N,N*-dimethyltryptamine (DMT), diisopropyltryptamine (DiPT)
1B. With a modification on the 4-position on the indole ring	Psilocin, psilocybin
1C. With a modification on the 5-position on the indole ring	Bufotenine, 5-methoxy-α-methyltryptamine (5-MeO-AMT), 5-methoxy-*N,N*-dimethyltryptamine (5-MeO-DMT), 5-methoxydiisopropyltryptamine (5-MeO-DiPT)
2. *Lysergamides*	Lysergic acid diethylamide (LSD), lysergic acid 2,4-dimethylazetidide (LSZ), 1-propionyl-D-lysergic acid diethylamide (1P-LSD), lysergic acid amide (LSA)
Phenethylamines	
1. *2,5-dimethoxyphenethylamines (2C-compounds)*	2,5-Dimethoxy-4-chlorophenethylamine (2C-C), 2,5-dimethoxy-4-iodophenethylamine (2C-I), 2,5-dimethoxy-4-methylthiophenethylamine (2C-T)
2. *N-(methoxybenzyl)-2,5-dimethoxyphenethylamines (NBOMe; 25-NBOMe compounds)*	4-Bromo-*N*-(methoxybenzyl)-2,5-dimethoxyphenethylamine (25B-NBOMe), 4-iodo-*N*-(methoxybenzyl)-2,5-dimethoxyphenethylamine (25I-NBOMe), 4-chloro-*N*-(methoxybenzyl)-2,5-dimethoxyphenethylamine (25C-NBOMe)
3. Dibenzodifurans	2C-B-FLY, Bromo-Dragonfly
Hallucinogens **B. Dissociatives**	
Arylcyclohexylamines	Methoxetamine (MXE), 2-methoxyketamine (2-MK), 4-methoxyphencyclidine (4-MeO-PCP)
Diarylethylamines	Ephenidine, diphenidine, methoxyphenidine
Benzodiazepines	
Phenazepam, pyrazolam, flubromazepam, *iso*-flubromazepam	
Opioids	
Fentanyl analogues	Acetylfentanyl, acryloyfentanyl, 3-methylfentanyl, butyrylfentanyl, 4-fluorobutyrylfentanyl, ocfentanil
Other	MT-45, AH-7921, U-47700

25,000 seizures (1.8 tonnes) of designer cathinones were reported in Europe (EMCDDA 2017). On the global scale, in 2014 seizures reached the level of 32 and 1.3 tonnes for synthetic cannabimimetics and cathinones, respectively (UNODC 2016).

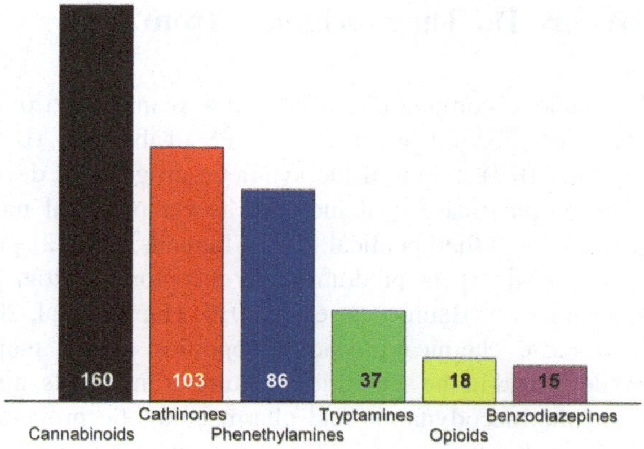

Fig. 2.2 Number of compounds from the main NPS groups notified to the European Monitoring Centre for Drugs and Drugs Addiction by the end of 2015. (EMCDDA 2013b)

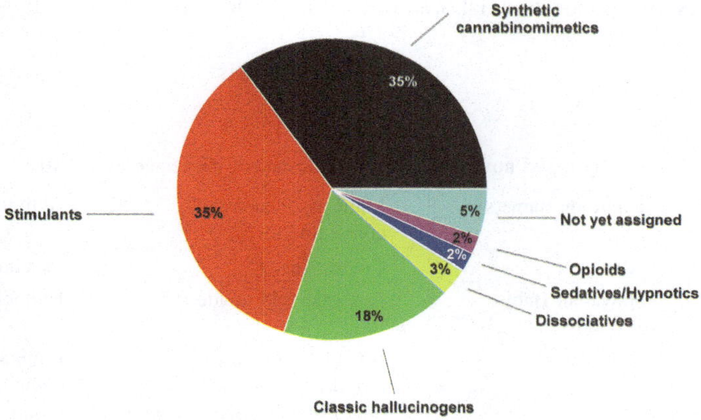

Fig. 2.3 Proportion of different NPS groups in the global sale by December 2015 (UNODC 2016)

In recent years, a range of different synthetic opioids, namely, analogues of fentanyl and compounds with various chemical structures (AH-7921, U-47700 and MT-45), has appeared in both Europe and North America. As these highly potent compounds are often mixed up with heroin ('fake heroin') to masquerade heroin, are included in cocaine products, or are pressed into counterfeit prescription pills, their presence on the illicit drug market is causing considerable concern (Zawilska 2017).

2.4 NPS—Where Do They Originate from?

NPS are mostly synthetic compounds. Only a few plants/fungi or their extracted products have been categorized and used as NPS (Table 2.2) (Baumeister et al. 2015; Graziano et al. 2017). Some of the synthetic drugs were developed by academic laboratories or pharmaceutical industry as (1) potential medicines which were never approved for a therapeutical use in humans, and (2) pharmacological tools to study biochemical targets, predominantly receptors and transporter proteins, for brain neurotransmission (Baumeister et al. 2015; Huestis et al. 2017). Although the synthesis and basic chemical/physical properties of the majority of these compounds were described in the scientific literature or in patents, almost nothing is known about their pharmacodynamic and pharmacokinetic properties in humans, and their potential for abuse. Several NPS are derivatives of controlled substances whose chemical structures have been slightly modified. It should be emphasized that even small molecular differences from previously controlled compounds can result in a considerable difference in terms of biological activity, pharmacokinetic parameters and identification. Other substances have been created as entirely new compounds by clandestine laboratories (Baumeister et al. 2015; Huestis et al. 2017).

Table 2.2 Plants/fungi-based novel psychoactive substances (Graziano et al. 2017)

Species/genera	Common name(s)	Main psychoactive compound(s)	Activity profile
Catha edulis	Khat	Cathinone	Psychostimulant
Argyreia nervosa	Hawaiian Baby Woodrose	Lysergic acid amide (LSA)	Hallucinogenic
Piper methysticum	Kava-kava	Kavalactones (kavain, dihydrokavain, methysticin, dihydromethysticin, yangonin and desmethoxyyangonin)	Euphoric, entheogenic, sedative, anxiolytic
Mitragyna speciosa Korth	Kratom	Mitragynine, 7-hydroxymitragynine	Opioid-like
Salvia divinorum		Salvinorin A	Hallucinogenic
Psychodria viridis, *Mimosa hostilis*		*N,N*-Dimethyltryptamine	Hallucinogenic
Psilocybe	Magic mushrooms, psilocybin mushrooms, psychedelic mushrooms	Psilocin and psilocybin	Hallucinogenic
Tabernanthe iboga		Ibogaine	Hallucinogenic

2.5 Why Are NPS Attractive to Their Users?

NPS affect every social groups, but are popular among socially marginalized groups, drug addicts, rave parties participants and members of online communities. There are several motivating factors for NPS use (see also Chap. 9). In general, they can be divided into two distinct groups (Deligianni et al. 2017; Soussan and Kjellgren 2016; Sutherland et al. 2017; for more details see Chap. 9):

(1) External—price, legal status, availability and non-detectability in screening tests;
(2) Internal—curiosity, enjoyable effects and/or enhancement of social situations.

A common motivation for using any NPS is price, pleasure and enjoyment. The fact that currently no accurate field testing devices exist for most NPS also attracts many potential users. The most highly endorsed motivation for the use of synthetic cathinones and cannabimimetics is availability, and for the latter group, their legality (or perceived legality). Stimulants are typically used to enhance psychophysical activity, while the use of hallucinogens is highly motivated by exploration and spiritual attainment (Soussan and Kjellgren 2016). Last but not least, the growing popularity of NPS would not be possible without the Internet. Several websites, discussion fora and blogs constitute an important platform to share information about reliable online shops, offered products and their prices, newer compounds, the effects of drugs and harm reduction actions (Abdulrahim and Bowden-Jones 2015; EMCDDA 2016b; European Commission 2014; Huestis et al. 2017; Wadsworth et al. 2017; for more details see Chap. 9).

2.6 Pharmacological Properties of NPS

Despite the popularity of NPS, little is known about their metabolism and interaction with other xenobiotics, including medicines. Some of them, e.g. synthetic cannabiminetics and cathinones, could undergo extensive biotransformation by the liver microsomal enzymes, leading to both biologically active and inactive metabolites, with a concomitant very low or even negligible levels of the parent compound in examined biological fluids. This situation emphasizes the urgent need for a development of highly sensitive analytical methods for the detection and quantification of NPS metabolic constituents. Furthermore, greater knowledge of the specific enzymes involved in the biotransformation of particular compounds will improve our ability to predict potential drug–drug interactions and the possible impact of genetic polymorphism on idiosyncratic toxicity. There is also a paucity of scientific data on molecular mechanisms of action of these drugs. Accumulating experimental evidence indicates that NPS interact with various neurotransmitter targets (Baumeister et al. 2015; Logan et al. 2017; Schifano et al. 2017). Psychostimulants elevate the extracellular level of monoamines in the brain by selective and very potent inhibition of dopamine and noradrenaline reuptake by the inhibition of DAT and NET transporters, respectively (see Chap. 6). Some of them (methamphetamine-like cathinones) act as monoamine transporter substrates and stimulate neurotransmitter release. Those additionally endowed with emphatic/entactogenic activity (i.e. phenylpiperazines, aminoindanes and MDMA-like cathinones) predominantly act at serotonin transporter (SERT) (Liechti 2015). Noteworthy, a high DAT/SERT inhibition ratio is a pharmacological characteristic associated with more pronounced psychostimulant effects and with a higher abuse potential (Liechti 2015). Synthetic cannabimimetics exert their psychoactive effects by stimulating cannabinoid type 1 (CB_1) receptors (Castaneto et al. 2014; Zawilska and Wojcieszak 2014). Most psychedelic compounds are agonists or partial agonists of serotonin receptors (predominantly 5-HT_{2A}; some of them also stimulate 5-HT_{1A}, 5-HT_{2C} and 5-HT_{2B} receptors). Among psychedelics, one can also find compounds that activate sigma1 receptors, central M_1-cholinergic receptors, or, like salvinorin A, κ-opioid receptors (Tittarelli et al. 2015; Zawilska and Andrzejczak 2015; Zawilska and Wojcieszak 2013). Dissociatives primarily act as non-competitive antagonists of glutamatergic ionotropic NMDA receptors. In addition, they variously interact with serotonin, opioid, adrenergic, cholinergic and dopamine receptors (Baumeister et al. 2015; Beharry and Gibbons 2016). Benzodiazepines increase GABA-ergic neurotransmission by binding to an allosteric modifying site on the $GABA_A$ receptor complex (Baumeister et al. 2015). Synthetic opioids act as agonists of opioid receptors, predominantly the μ-, and to a lesser extent, the δ-subtype (Zawilska 2017).

2.7 NPS Exert Various Adverse Effects

The use of NPS is associated with a wide spectrum of unwanted effects, the knowledge of which is derived almost exclusively from occasional case reports of acute intoxication and fatal poisonings. Some adverse effects are characteristic of particular groups of compounds: for example, constricted pupils, respiratory depression and cyanosis are indicative of intoxication with synthetic opioids (Zawilska 2017), while numerous symptoms could be produced by drugs from different NPS classes. In addition, there are important individual variations in the response to NPS and similar doses may have significantly different toxic effects and consequences in different individuals (Logan et al. 2017).

The most commonly reported adverse effects of NPS include (Logan et al. 2017):

Psychotic—psychosis with delusions, dysphoria, agitation, aggression that progresses sometimes to violent or even criminal behaviour and self-destructive behaviour, irritability, anxiety and panic attacks. Concerns are raised that the use of at least some types of NPS (cannabimimetics, psychostimulants, hallucinogens and psychodysleptics) may exacerbate existing psychiatric symptoms.

Neurological—drowsiness, headache, motor incoordination, seizures, slurred speech, disturbed sleep patterns and nightmares, insomnia and blurred vision.

Cognitive—confusion, short-term memory deficits, cognitive impairment, mental fatigue, disorientation to name, place and time, loosening of association and disorganized thoughts.

Cardiovascular—tachycardia, hypertension, palpitation, chest pain, acute myocardial infarction and cerebral ischemia.

Miscellaneous—nausea, vomiting, diaphoresis, disturbances in the acid–base balance and electrolyte balance, and acute kidney injury.

In severe cases, NPS led to disseminated intravascular coagulation, rhabdomyolysis, cardiac arrest, respiratory depression or multiorgan failure. Lethal effects were reported after both accidental and deliberate overdose of these compounds (Abdulrahim and Bowden-Jones 2015; Logan et al. 2017).

With the exception of opioids and benzodiazepines, there are no specific antidotes for NPS. Clinical management is generally orientated towards providing symptomatic and supportive care, with a special attention to cardiac and respiratory functions. Often it may require admission to the intensive care units. Renal, muscular and hepatic functions should be also monitored. Benzodiazepines (e.g. lorazepam, diazepam or midazolam) are the first-line pharmacological treatment for the agitation, tremor and seizures, while aggression and psychosis may require the use of antipsychotic drugs (e.g. risperidone, ziprasidone or haloperidol). Parenteral sedation with lorazepam or propofol is used in cases of extreme agitation. Hyperthermia and serotonin syndrome are managed by cyproheptadine administration in addition to cooling and benzodiazepines. When body temperature is higher than 41 °C, immediate paralysis with non-depolarizing agents, tracheal intubation and ventilation are recommended. Antipyretics and dantrolene are ineffective. Treatment of hyponatremia in asymptomatic patients requires fluid restriction, while sodium correction with hypertonic saline is recommended in patients with neurological symptoms (Abdulrahim and Bowden-Jones 2015).

2.8 Concluding Remarks

During the last decade, NPS have extensively dominated the drug scene in different parts of the world. In 2016, Commission on Narcotic Drugs expressed its deep concern '..about the exploitation or misuse of the Internet and social media by drug traffickers in the sales and promotion of illicit drugs, new psychoactive substances and precursors, as well as the increasing use of communication technologies, including the Internet, online payment systems and virtual currencies, for purchasing these substances.' (Commission on Narcotic Drugs 2016).

Although the prevalence and pattern of NPS use differ between countries, the most popular groups are synthetic cannabimimetics and psychostimulatory designer cathinones, the latter being the subject of the present book. Currently, there is limited information available on the potential acute toxicity (harms) associated with the use of these substances. However, the number of intoxicated people presenting with emergencies is constantly increasing, indicating that recreational and chronic users may indeed be subject to negative health and social consequences to a serious degree. Considering the ever-changing market of NPS, both on the local and global scale, we need to raise awareness of these compounds, and to substantially facilitate information sharing regarding recreational drugs trends, experience and toxicity. As emphasized by the Commission on Narcotic Drugs, there is a need for further research into and development of scientific evidence-based responses to help those providing healthcare and social services to prevent and address the adverse health consequences associated with NPS use (Commission on Narcotic Drugs 2017).

Acknowledgements This research was supported by the National Science Centre (NCN), Cracow, Poland (Grant No. 2014/13/B/NZ7/02237).

References

Abdulrahim D, Bowden-Jones O, on behalf of the NEPTUNE Expert Group (2015) Guidance on the management of acute and chronic harms of club drugs and novel psychoactive substances. Novel Psychoactive Treatment UK Network (NEPTUNE), London. http://neptune-clinical-guidance.co.uk/wp-content/uploads/2015/03/NEPTUNE-Guidance-March-2015.pdf. Accessed 24 Apr 2017

Baumeister D, Tojo LM, Tracy DK (2015) Legal highs: staying on the top of the flood of novel psychoactive substances. Ther Adv Psychopharmacol 5(2):97–132. https://doi.org/10.1177/2045125314559539

Beharry S, Gibbons S (2016) An overview of emerging and new psychoactive substances in the United Kingdom. Forensic Sci Int 267:25–34. https://doi.org/10.1016/j.forsciint.2016.08.013

Brunt TM, Poortman A, Niesink RJ et al (2011) Instability of the ecstasy market and a new kid on the block: mephedrone. J Psychopharmacol 25(11):1543–1547. https://doi.org/10.1177/0269881110378370

Castaneto MS, Gorelick DA, Desrosiers NA et al (2014) Synthetic cannabinoids: epidemiology, pharmacodynamics, and clinical implications. Drug Alcohol Depend 144:12–41. https://doi.org/10.1016/j.drugalcdep.2014.08.005

Commission on Narcotic Drugs Vienna (2016) Resolution 59/8. Promotion of measures to target new psychoactive substances and amphetamine-type stimulants. https://www.unodc.org/documents/commissions/CND/CND_Sessions/CND_59/Resolution_59_8.pdf. Accessed 6 Nov 2017

Commission on Narcotic Drugs Vienna (2017) Resolution 60/4. Preventing and responding to the adverse health consequences and risks associated with the use of new psychoactive substances. https://www.unodc.org/documents/commissions/CND/CND_Sessions/CND_60/CNDres_2017/Resolution_60_4_60CND.pdf. Accessed 6 Nov 2017

Corazza O, Valeriani G, Bersani FS et al (2014) "Spice," "kryptonite," "black mamba": an overview of brand names and marketing strategies of novel psychoactive substances on the web. J Psychoactive Drugs 46(4):287–294. https://doi.org/10.1080/02791072.2014.944291

Deligianni E, Corkery JM, Schifano F et al (2017) An international survey on the awareness, use, preference, and health perception of novel psychoactive substances (NPS). Hum Psychopharmacol 32(3). https://doi.org/10.1002/hup.2581

European Commission (2014) Flash Eurobarometer 401: young people and drugs report. European Commission, Brussels. http://ec.europa.eu/commfrontoffice/publicopinion/flash/fl_401_en.pdf. Accessed 18 July 2017

European Monitoring Centre for Drugs and Drug Addiction (2012) Annual report 2012. The state of the drug problem in Europe. http://www.emcdda.europa.eu/attachments.cfm./att_190854_EN_TDAC13001ENC_pdf. Accessed 9 June 2014

European Monitoring Centre for Drugs and Drug Addiction (2013) European Drug Report 2013: trends and developments. http://www.emcdda.europa.eu/system/files/publications/964/TDAT13001ENN1.pdf. Accessed 9 June 2014

European Monitoring Centre for Drugs and Drug Addiction (2015a) New psychoactive substances in Europe. An update from the EU Early Warning System (March 2015). http://www.emcdda.europa.eu/attachements.cfm/att_235958_EN_TD0415135ENN.pdf. Accessed 23 Apr 2015

European Monitoring Centre for Drugs and Drug Addiction (2015b) Perspective on drugs. Synthetic cannabinoids in Europe. http://www.emcdda.europa.eu/topics/pods/synthetic-cannabinoids. Accessed 25 July 2015

European Monitoring Centre for Drugs and Drug Addiction (2016a) Health responses to new psychoactive substances. http://www.emcdda.europa.eu/system/files/publications/2155/TDXD16001ENN_FINAL.pdf. Accessed 30 Mar 2017

European Monitoring Centre for Drugs and Drug Addiction (2016b) European Drug Report 2016: trends and developments. http://www.emcdda.europa.eu/system/files/publications/2637/TDAT16001ENN.pdf. Accessed 30 Mar 2017

European Monitoring Centre for Drugs and Drug Addiction (2016c) EU Drug Market Report. In-depth analysis. http://www.emcdda.europa.eu/system/files/publications/2373/TD0216072ENN.PDF. Accessed 15 Feb 2017

European Monitoring Centre for Drugs and Drug Addiction (2016d) The internet and drug markets. http://www.emcdda.europa.eu/system/files/publications/2155/TDXD16001ENN_FINAL.pdf. Accessed 15 May 2017

European Monitoring Centre for Drugs and Drug Addiction (2017) European Drug Report 2017: trends and developments. http://www.emcdda.europa.eu/system/files/publications/4541/TDAT17001ENN.pdf. Accessed 21 June 2017

Guirguis A, Corkery JM, Stair JL et al (2017) Intended and unintended use of cathinone mixtures. Hum Psychopharmacol 32(3). https://doi.org/10.1002/hup.2598

Graziano S, Orsolini L, Rotolo MC et al (2017) Herbal highs: review on psychoactive effects and neuropharmacology. Curr Neuropharmacol 15:750–761. https://doi.org/10.2174/1570159X14666161031144427

Helander A, Bäckberg M (2017) New psychoactive substances (NPS)—the Hydra monster of recreational drugs. Clin Toxicol (Phila) 55:1–3. https://doi.org/10.1080/15563650.2016.1217003

Huestis MA, Brandt SD, Rana S et al (2017) Impact of novel psychoactive substances on clinical and forensic toxicology and global public health. Clin Chem 63(10):1564–1569. https://doi.org/10.1373/clinchem.2017.274662

Lechti ME (2015) Novel psychoactive substances (designer drugs): overview and pharmacology of modulators of monoamine signaling. Swiss Med Wkly 145:w14043

Logan BK, Mohr ALA, Friscia M et al (2017) Reports of adverse events associated with use of novel psychoactive substances, 2013–2016: a review. Anal Toxicol 41(7):573–610. https://doi.org/10.1093/jat/bkx031

Orsolini L, Francesconi G, Papanti D et al (2015) Profiling online recreational/prescription drugs' customers and overview of drug vending virtual marketplaces. Hum Psychopharmacol 30(4):302–318. https://doi.org/10.1002/hup.2466

Papaseit E, Farré M, Schifano F et al (2014) Emerging drugs in Europe. Curr Opin Psychiatry 27(4):243–250. https://doi.org/10.1097/YCO.0000000000000071

Schifano F, Orsolini L, Papanti D et al (2017) NPS: medical consequences associated with their intake. Curr Topics Behav Neurosci 32:351–380. https://doi.org/10.1007/7854_2016_15

Soussan C, Kjellgren A (2016) The users of novel psychoactive substances: online survey about their characteristics, attitudes and motivations. Int J Drug Policy 32:77–84. https://doi.org/10.1016/j.drugpo.2016.03.007

Sutherland R, Bruno R, Peacock A et al (2017) Motivations for new psychoactive substance use among regular psychostimulant users in Australia. Int J Drug Policy 43:23–32. https://doi.org/10.1016/j.drugpo.2016.12.021

Tittarelli R, Mannocchi G, Pantano F et al (2015) Recreational use, analysis and toxicity of tryptamines. Curr Neuropharmacol 13(1):26–46. https://doi.org/10.2174/1570159X13666141210222409

United Nations Office on Drugs and Crime (2013a) Worlds Drug Report 2013. New psychoactive substances: concepts and definitions, Vienna, pp. 60–115. http://www.unodc.org/unodc/secured/wdr/wdr2013/World_Drug_Report_2013.pdf. Accessed 22 Sept 2014

United Nations Office on Drugs and Crime (2013b) The challenge of new psychoactive substances. A report from the global SMART programme, March 2013. https://www.unodc.org/documents/scientific/NPS_Report.pdf. Accessed 18 July 2017

United Nations Office on Drugs and Crime (2016) World Drug Report 2016. http://www.unodc.org/doc/wdr2016/WORLD_DRUG_REPORT_2016_web.pdf. Accessed 17 July 2017

Vandrey R, Johnson MW, Johnson PS et al (2013) Novel drugs of abuse: a snapshot of an evolving marketplace. Adolesc Psychiatry (Hilversum) 3(2):123–134

Wadsworth E, Drummond C, Kimergård A, Deluca P (2017) A market on both "sides" of the law: the use of the hidden web for sale of new psychoactive substances. Hum Psychopharmacol. https://doi.org/10.1002/hup.2596

Zawilska JB (2011) "Legal highs"—new players in the old drama. Curr Drug Abuse Rev 4 (2):122–130

Zawilska JB (2017) An expanding world of novel psychoactive substances: opioids. Front Psych 8. https://doi.org/10.3389/fpsyt.2017.00110

Zawilska JB, Andrzejczak D (2015) Next generation of novel psychoactive substances on the horizon—a complex problem to face. Drug Alcohol Depend 157:1–17. https://doi.org/10.1016/j.drugalcdep.2015.09.030

Zawilska JB, Wojcieszak J (2013) *Salvia divinorum*: from Mazatec medicinal and hallucinogenic plant to emerging recreational drug. Hum Psychopharmacol 28(5):403–412. https://doi.org/10.1002/hup.2304

Zawilska JB, Wojcieszak J (2014) Spice/K2 drugs—more than innocent substitutes for marijuana. Int J Neuropsychopharmacol 17(3):509–525. https://doi.org/10.1017/S1461145713001247

Zuba D, Byrska B (2013) Prevalence and co-existence of active components of 'legal highs'. Drug Test Anal 5(6):420–429. https://doi.org/10.1002/dta.1365

Zuba D, Byrska B, Maciow M (2011) Comparison of "herbal highs" composition. Anal Bioanal Chem 400(1):119–126. https://doi.org/10.1007/s00216-011-4743-7

Chapter 3
Khat—A Natural Source of Cathinone

Nilesh B. Patel

Abstract Recently, a class of new psychoactive substances (NPS) has appeared as drugs of abuse. NPS comprise different drug classes; the most popular are synthetic cannabinomimetics and designer analogues of cathinone, an intermediate metabolite in a biosynthetic pathway, and a psychostimulant, found in the plants *Catha edulis* Forsk., *Ephedra gerardiana sikkimensis*, and *Ephedra sinica*. In *C. edulis* and *E. gerardiana sikkimensis*, cathinone is converted to norpseudoephedrine (cathine) and norephedrine, while in *E. sinica* the pathway continues to pseudoephedrine and ephedrine. While cathinone analogues are used for recreational purposes, cathinone, itself, is consumed only through chewing khat, i.e., the fresh leaves and twigs of the plant *C. edulis* Forsk., a practice similar to chewing of coca leaves. The use of khat is common among people living in Southwestern Arabia and Eastern Africa with an estimated 10–20 million daily users. This chapter describes the use of khat, its distribution and cultivation, the biosynthetic pathway that involves cathinone, the long delay in the discovery of cathinone, the neurochemistry of khat and cathinone, its effects on humans, and its dependency and addiction potential.

Keywords Khat · Cathinone · Novel psychoactive substances
Catha edulis Forsk · *Ephedra gerardiana sikkimensis* · *Ephedra sinica*

3.1 Khat and Its Consumption

Khat, the young fresh leaves and twigs of *Catha edulis* Forsk. (*Celastraceae*) is usually chewed in a group or social setting for psychostimulatory effects. It may also be consumed as a beverage—Abyssinian, Arabic, or Bushman's tea. Its use serves to lubricate and enhance social discourse and cultural and traditional

N. B. Patel (✉)
Department of Medical Physiology, School of Medicine, College of Health Sciences, University of Nairobi, P. O. Box 30197-00100, Nairobi, Kenya
e-mail: npatel@uonbi.ac.ke

J. B. Zawilska (ed.), *Synthetic Cathinones*, Current Topics in Neurotoxicity 12,
https://doi.org/10.1007/978-3-319-78707-7_3

celebrations. In Yemen, khat was introduced from Ethiopia in the middle of the fifteenth century, and the practice of khat chewing is regulated by traditional and social norms rather than legal mechanisms. Here, it has become a deeply rooted cultural and traditional practice and acquired a social status. There are formal chewing sessions, e.g., after afternoon or evening prayers, and many homes have a room set aside for khat-chewing sessions. In other countries, khat use is as the person wishes; there is no social, cultural, or legal limitation on its use.

Most of the reports on the practice and habit of khat chewing are based on observations of its use in Yemen and Jazan region of Saudi Arabia (though it is illegal in Saudi Arabia). To chew khat, the leaves and bark from the fresh twigs are removed and stored as a wad or quid in the cheek of the mouth and masticated until the "juice" is extracted and swallowed. This takes about 10–15 min. The used up khat material is then discarded, or in some cases swallowed, and more fresh leaves and twigs are chewed. To reduce the astringent taste and prevent dryness in the mouth, sweet beverages and drinks are consumed, and chewing gum is also used to bind the material together. A khat-chewing session can last 3–4 h and between 100 and 400/500 g can be consumed or, at times, the users engage in binge sessions lasting 24 or more hours (Klein and Beckerleg 2007; Nabuzoka and Badhadhe 2000; Odenwald et al. 2007; Widmann et al. 2014; Patel et al. 2005).

The fresh khat leaves contain different compounds, including alkaloids (majority of which are phenethylamines and cathedulins), glycosides, tannins, flavonoids, terpenoids, amino acids, vitamins, and minerals (Szendrei 1980). The khat phenethylamines comprise cathinone (*S*-(−)-cathinone), and two diastereoisomers: (+)-norpseudoephedrine 1*S*,2*S*-(+)-norpseudoephedrine; cathine) and (−)-norephedrine (1*R*,2*S*-(−)-norephedrine). The preference of the user for fresh khat leaves and twigs was only understood after the discovery of cathinone, the major psychoactive compound of the plant (described in a later section). Analysis of the whole khat plant by Krizevski et al. (2007) showed that cathinone has a juvenile distribution pattern. It is found in the highest concentration in the young leaves, twigs, and flowers, while older plant parts have high levels of cathine and norephedrine and low levels of cathinone.

The amount of cathinone in khat varies depending on the source and cultivation. Analysis of fresh khat from Ethiopia, Kenya, North Yemen, and Madagascar showed 0.9–3.3‰ cathinone, 0.05–7.5‰ cathine, and 0.007–0.8‰ norephedrine (Geisshusler and Brenneisen 1987). From this, in a single khat session of 4 h around 0.09–1.32 g of cathinone could be consumed at a rate of around 22–330 mg/h over a 4 h session.

Plasma cathinone, cathine, and norephedrine levels during a regular khat chewing have not been measured. In volunteers chewing fresh khat for 1 h (average chewed between 44 and 60 g/subject), peak plasma cathinone concentration was reached after 1.5–3.5 h, and the maximum plasma cathinone level measured was 59–127 μg/L with a terminal elimination half-life between 90 and 260 min (Toennes et al. 2003; Widler et al. 1994).

Toennes et al. (2003) also studied the pharmacokinetics of the alkaloids cathinone, cathine, and norephedrine from chewing fresh khat leaves and proposed a two-compartment model with two-segment absorption. The residue of khat chewed for 1 h had 90% of its alkaloids released, and the buccal mucosa played a major role in the alkaloids' absorption, while the stomach or small intestine or both tissues were responsible for the second phase of absorption. The alkalinity of human salvia also increased the extraction of cathinone (Krizevski et al. 2007). In another study, using Caco-2 cells (derived from human colon carcinoma) monolayer cultures, porcine buccal and sublingual mucosal tissue, and intestinal epithelial tissue, Atlabachew et al. (2016) found that permeation of the three main psychoactive alkaloids of khat, i.e., cathinone, cathine, and norephedrine, was as good as the permeation of the reference substance caffeine. In each case, cathinone permeation across the monolayer of Caco-2 cells and porcine tissues was higher than of cathine and norephedrine, and both the oral mucosal and intestinal tissue had good permeation.

3.2 Distribution and Cultivation of Khat

C. edulis Forsk., a hardy perennial dicotyledonous evergreen shrub, cultivated as a bush or a tree, is a natural source of the psychostimulant cathinone, the analogues of which made up a part of the New Psychoactive Substances (NPS) list. It is a member of the *Celastraceae* family and usually grows at altitudes of 1500–2000 m in humid mountain/hill regions and can reach heights of 6 m. It was given its scientific name by Peter Forskal, and published in the 1775 botanical papers of Karsten Niebuhr. Khat grows in Southwestern Arabian Peninsula, and down south through Eastern Africa, Madagascar, and Southern African, and west to Congo; it is also cultivated in Indonesia. The major commercial cultivation areas of khat are the Harar Province, Ethiopia; Sanaa-Taiz, Yemen; Nyambene area of Meru, Kenya, and Cianjur, Indonesia. Depending on soil conditions, weather, and location, more than 40 different varieties of khat are described (Zahran et al. 2014). Some of the local names of khat are gat (Yemen); qaat, gat, jaad (Somalia); dimma, hamecot, chat (Ethiopia); and miraa (Kenya) (Feyissaa and Kelly 2008).

Cultivation is mainly done by use of cuttings, and the plant is ready for harvesting in 3–4 years. Each tree can be harvested 2–3 times a year. Harvesting involves plucking the young leaves and twigs which are then packed into bundles of 40 or so, wrapped in leaves, e.g., banana leaves, or put into plastic bags, to keep them moist, and transported to the markets where they can be sold for up to 3–4 days post-harvest, after which they lose their potency to induce psychostimulatory and euphorigenic effects in the user (Cox and Rampes 2003). Due to its perishable nature, until recently, khat use was confined to regions where the plant was grown. However, with the advent of modern rapid refrigerated transport, and the emigration of people from the traditionally khat using countries, e.g., Yemen, Ethiopia, and Somalia, the distribution and use of fresh khat has spread to other parts of the world, and a number of countries have imposed a ban on its importation and consumption (en.wikepedia.org/wiki/khat 2017).

While the number of persons who consume khat on a regular basis is not known, it is estimated to be around 10–20 million (Al-Motarreb et al. 2002; Magdum 2011; Saha and Dollery 2006). This usage figure is comparable to other psychostimulants reported in 2015 World Drug Report—cocaine: 13.8–20.7 million; amphetamine: 13.8–53.8 million; and ecstasy: 9.3–28.4 million (UNODC 2015). In countries where khat use is legal and culturally acceptable, daily khat use in adult male population ranges from 6 to 75% and for females around 10% (see review by Patel 2015). Lifetime prevalence can be very high, e.g., over 90% in Yemen where in 2008 WHO reported that 15–20% of children under the age of 12 years use khat (WHO 2008).

There are a couple of striking characteristics of khat users. One, the majority are Muslims and the highest prevalence of khat use is in Muslim countries (Yemen, Somalia, and Djibouti) or in countries with large Muslim population, like Kenya and Ethiopia. The preference for Muslims to use khat is based whether it is permissible (halal) or not (haram) under the Muslim Sharia Law (Hersi and Abdalla 2013), and opinion among the Muslim scholars is divided; hence, in some Muslim countries, it is prohibited while in others it is not. Several books describing this unique practice have been published (Anderson et al. 2007; Gebissa 2004). The second feature, which derives from the acceptance by Muslims of the khat use, is the majority of users do not use any other drugs, except perhaps tobacco, as alcohol and other intoxicants or psychoactive agents are prohibited by their religion. This probably makes khat users the largest global population of users of only one type of psychoactive compound.

3.3 Identification and Isolation of Cathinone

Between the fourteenth and eighteenth centuries, the European Age of Exploration, new worlds, peoples, and cultures were discovered by the Europeans. Among the many reports from these new places that were reaching Europe, there were reports of plants with psychostimulant properties. Visitors and travelers to the Arabian region sent commentaries and reports to their homelands on the use of khat by the people in Yemen and Ethiopia. For example, James Vaughan, the Port-Surgeon, Aden, Yemen, wrote his observations on the use of khat in "Notes upon the drugs observed at Aden, Arabia" to the London Pharmaceutical Journal in 1859 (see Hunter 1968). There were also commentaries and reports reaching Europe from Central and South America about another plant, *Erythroxylum coca*, the leaves of which also produced psychostimulatory effects. These reports stimulated the interests of the Europeans chemists and pharmacologists.

In 1855, Friedreich Gaedcke isolated the alkaloid, cocaine, and its structure was elucidated by Richard Willstatter in 1898. Cocaine gained popularity in medical treatment, especially as a local anesthetic in ophthalmology; it provided a way of conducting eye surgery without using general anesthesia (Goerig et al. 2012).

Its use then moved into social and recreational use. One of the proponents of its wonderful effect was Sigmund Freud who wrote on its "…exhilaration and lasting euphoria…".

The Chinese herb Ma-huang (Ephedra), which was listed in the pharmacopeia of the Chinese Emperor, Shen Nung (~2700 BC), was known for its stimulatory effects. In 1885, Nagayoshi Nagai in Japan isolated and determined the structure of the active molecule—ephedrine [(1*R*,2*S*)-2-(methylamino)-1-phenylpropan-1-ol]. In the 1920s, ephedrine was rediscovered and had a meteoric rise as a treatment for asthma. Lazar Edeleanu in 1887 synthesized amphetamine (deoxy-norephedrine), an analogue of ephedrine, and Nagayoshi Nagai in 1893 synthesized methamphetamine. Ephedrine and its analogues by the 1930s had reached enormous commercial success as medicinal agents (Appendino et al. 2014; Lee 2011), and also had migrated to social and recreational use. But while in the nineteenth century, chemists were having success in isolating and identifying the active agents in *Ephedra sinica* and *Erythroxylum coca*, the isolation and identification of psychostimulatory molecules of *C. edulis* Forsk. proved difficult due to the limits of the isolation techniques and knowledge of structural chemistry available to medicinal chemists at that time. Contributing to the difficulty in isolating cathinone was that it is an intermediate metabolite, not a terminal product, in a biosynthetic pathway in *C. edulis*, and its concentration in the plant rapidly declines after harvesting. Therefore, having material with a sufficient cathinone content for analysis was a key to the discovery of cathinone. One wonders if khat's main psychostimulatory molecule, cathinone, been isolated during the same period when ephedrine and cocaine were discovered, it might have had a much earlier history of social and recreational use.

But once the stimulating effects of *C. edulis* became known in Europe, the search was on to discover its active molecule(s). In 1887, Fluchiger and Geroch isolated a basic material which they termed "katine". This was followed by Mosso in 1891, who also extracted a basic material but could not crystallize it. This material, "celastrine", showed stimulating effect on the frog heart and, also, had lethal effects on frogs. The presence of an extractable basic material, and that it caused paralysis, was also reported by Beitter (1901). The rediscovery of ephedrine in the 1920s enabled Wolfes (1930), using ephedrine as a model molecule, to identify dextro-norpseudoephedrine in the *C. edulis* material.

Following the reservation raised by von Brücke (1941), who had found from simple experiments that dextro-norephedrine was not potent enough to account for khat's psychostimulatory effects, Alles et al. (1961) reviewed the previous work on identification of the active substance in khat and found that their khat extraction did have the pressor effect, as shown by others, as well as producing subjective effects when the solution was drunk. However, aside from identifying dextro-norephedrine in their extraction, they did not identify any other molecule, especially one that had amphetamine-like molecular characteristics. In retrospect, their failure to identify another active molecule despite careful study was probably due to an unintentional error in the preparation and transport of the khat material from Ethiopia. As described by Alles et al. (1961), the khat material used for their study was "…dried in the tropical sun and air for two days … put into canvas bags and sent by plane in transit through Aden [Yemen], then by ship to London, and finally to Los Angeles". Despite having access to fresh khat growing in Miami, Florida, and Pasadena, California, they processed the khat from these sources in a manner similar to the khat they had obtained in Ethiopia. As described on the section on biosynthesis pathway in *C. edulis* Forsk., and the reason for the preference of the user for fresh khat, the cathinone content in fresh young khat leaves declines rapidly after harvesting due to its reduction to cathine (norpseudoephedrine) and norephedrine; hence, the khat material used by Alles et al. (1961) probably did not have high enough cathinone concentration to be identifiable by the techniques available to them.

This issue of differing reports on the main active molecule in khat led to the Commission on Narcotic Drugs (report of 24th session, 1975) recommendation that "In view of the indications of the presence in the fresh leaves of a labile precursor of cathine that would be active on the central nervous system (CNS) and of the consumers' preference for fresh material, it was considered essential that the plant material used in the studies be fresh or that its chemical composition be as close to that of the fresh material as possible" (see Szendrei 1980).

Following this recommendation, fresh khat material was collected from Kenya, Madagascar, and the Yemen Arab Republic, analyzed by the United Nations Narcotics Laboratory, and in 1978 Schorno and Steinegger reported the identification and absolute configuration of the main psychostimulatory molecule in khat: cathinone (α-phenylalkylamine, (−)-α-aminopropiophenone). It is part of the phenethylamine class of molecules, many of which are bioactive, and is a β-keto analogue of amphetamine. Further studies showed that cathinone represents nearly 70% of the alkaloid content in fresh leaves and the cathinone content varies between khat from different geographical sources (Schorno and Steinegger 1978; Schorno et al. 1982) as well as in the different parts of the plant (Guantai and Maitai 1982; Krizevski et al. 2007). So, nearly a century after the search for the active principle of khat started, the main psychostimulatory molecule was finally discovered and its chemical structure determined.

Ironically, cathinone and methcathinone, and other derivatives of D,L-ephedrine (aminopropiophenone) had been synthesized by Hyde et al. (1928). The pharmaceutical company, Parke Davis, in 1957, lodged a patent on these molecules. While cathinone did not enter social and recreational market, metcathinone did, especially in the former Soviet Union where it was used as an antidepressant in the 1920s and 1930s (Glennon et al. 1995), and then in 1989 it surfaced in Indiana, USA, when a student intern at Parke Davis came across the studies done by the company on the use of these ephedrine derivatives (Bailey 1995; Emerson and Cisek 1993).

3.4 Biosynthetic Pathway in *Catha edulis* Forsk

Among the plants that have been analyzed to isolate and identify the molecules with psychoactive properties have been the *Ephedra* species and *C. edulis* Forsk. In the *Ephedra* species, ephedrine, the main psychoactive molecule, initially proved to be very useful in the treatment of asthma, and later as a precursor molecule for the synthesis of derivatives or analogues for other medicinal and recreational use. Following the studies on the biosynthetic pathway in *Ephedra*, several studies have been done to elucidate the biosynthetic pathway in khat of its psychoactive molecules: cathinone, cathine, and norephedrine. Krizervski et al. (2007) found that cell-free extracts made from the young leaves and stems of *C. edulis* Forsk. showed the presence of an NADH-dependent enzymatic reduction of cathinone to (+)-cathine and (−)-norephedrine. Following this, Hagel et al. (2011) carried out expressed sequence tag analysis in *C. edulis* and *E. sinica* to identify putative enzymes involved in the different biochemical steps and their corresponding genes to elucidate the biosynthetic pathways in these two plants. Groves et al. (2015), using transcriptome profiling in these two plants, provided quite a complete picture of the biosynthetic steps. As shown in Fig. 3.1, in both plants L-phenylalanine is metabolized either via β-oxidation or non β-oxidation pathway to 1-phenylpropane-1,2-dione. This volatile molecule is then converted to an intermediate metabolite—cathinone. From cathinone, the final end products are formed: cathine and norephedrine in *C. edulis*, and pseudoephedrine and ephedrine in *E. sinica*. In *C. edulis*, the biochemical process terminates at cathine and norephedrine as the final methylation step to norephedrine and ephedrine is not present, unlike in the *E. sinica* (Hagel et al. 2011; Krizevki et al. 2007) (Fig. 3.2).

The elusiveness in finding cathinone as a precursor molecule in *Ephedra* was explained by Grue-Sorenson and Spenser (1994): "It is also evident from the results that reduction of cathinone to yield two noralkaloids [norpseudoephedrine and

Fig. 3.1 Fresh bundles of khat wrapped in banana leaves or newspaper (**a**). The khat on the left is "giza", the preferred variety of khat compared to "asili" on the right. **b** The separate bundles, and **c** the twigs from which the leaves and the bark is chewed

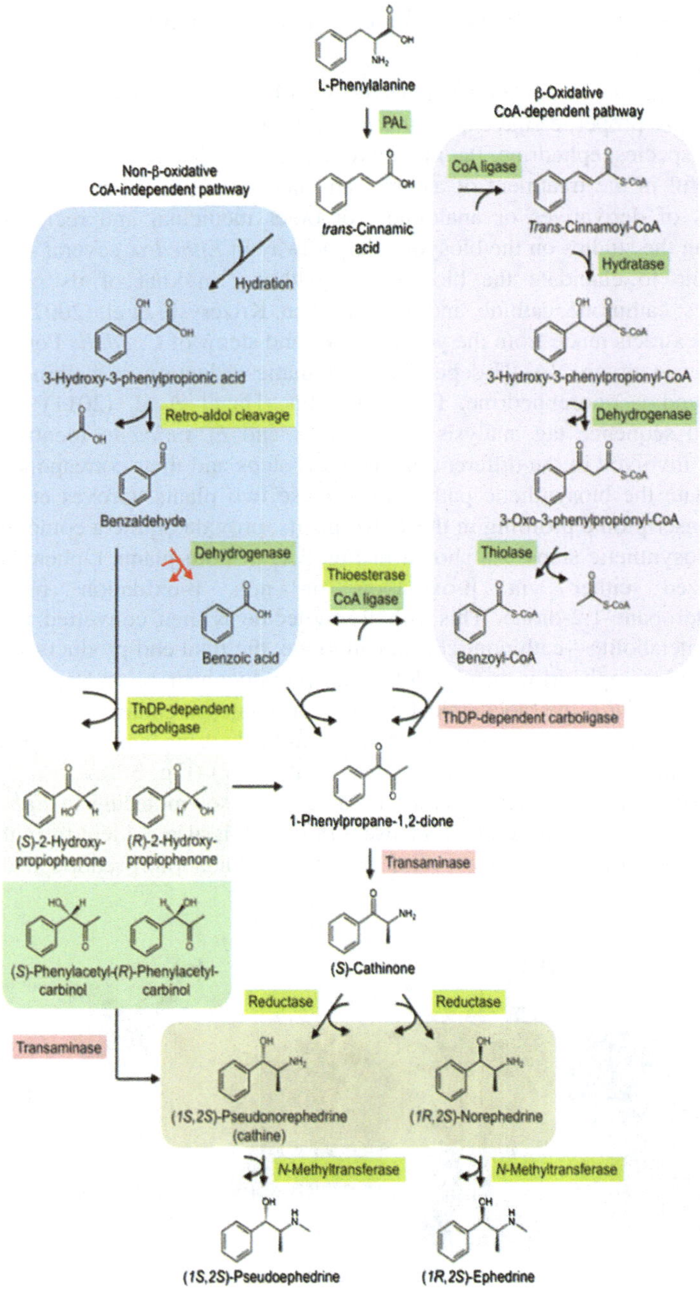
L-Phenylalanine
PAL
trans-Cinnamic acid
β-Oxidative CoA-dependent pathway
Non-β-oxidative CoA-independent pathway
CoA ligase
Trans-Cinnamoyl-CoA
Hydratase
Hydration
3-Hydroxy-3-phenylpropionic acid
3-Hydroxy-3-phenylpropionyl-CoA
Retro-aldol cleavage
Dehydrogenase
Benzaldehyde
3-Oxo-3-phenylpropionyl-CoA
Dehydrogenase
Thiolase
Thioesterase
CoA ligase
Benzoic acid
Benzoyl-CoA
ThDP-dependent carboligase
ThDP-dependent carboligase
(S)-2-Hydroxy-propiophenone
(R)-2-Hydroxy-propiophenone
1-Phenylpropane-1,2-dione
Transaminase
(S)-Phenylacetyl-carbinol
(R)-Phenylacetyl-carbinol
(S)-Cathinone
Reductase
Reductase
Transaminase
(1S,2S)-Pseudonorephedrine (cathine)
(1R,2S)-Norephedrine
N-Methyltransferase
N-Methyltransferase
(1S,2S)-Pseudoephedrine
(1R,2S)-Ephedrine

◂**Fig. 3.2** Proposed biosynthetic routes leading from L-phenylalanine to amphetamine-type alkaloids in *Catha edulis* and *Ephedra sinica*. A CoA-independent, non β-oxidative pathway of L-phenylalanine side chain shortening is shown in blue, whereas a CoA-dependent, β-oxidative route is shown in purple. Benzaldehyde, benzoic acid, and/or benzoyl-CoA undergo condensation with pyruvate, a reaction putatively catalyzed by a ThDP-dependent carboligase. 1-Phenylpropane-1,2-dione undergoes transamination to yield (S)-cathinone, which is reduced to cathine and (1*R*,1*S*)-norephedrine. *N*-methylation is restricted to *Ephedra* spp. and does not occur in khat. Activity has been detected for enzymes highlighted in yellow, and corresponding genes are available for enzymes highlighted in green. Enzymes highlighted in red have not been isolated, although database mining revealed numerous potential candidates. CoA Coenzyme A; NAD(H) nicotinamide adenine dinucleotide; NADP(H) nicotinamide adenine dinucleotide phosphate; PAL phenylalanine ammonia lyase; ThDP thiamine diphosphate. *Source* Groves et al. (2015)

norephedrine] in an almost 1:1 ratio is remarkably efficient. This facile transformation of cathinone into noralkaloids explains why cathinone had not hitherto been detected in or isolated from Ephedra species."

3.5 Neurochemistry of Cathinone

There are no studies done specifically on the neurochemical changes produced by khat. Once cathinone was identified, a number of studies were done on its neurochemistry, and it is assumed that the neurochemistry of cathinone would reflect that of khat. In the 1980s, a series of studies looked at cathinone's affect on the release of dopamine (DA), serotonin (5-HT), or noradrenaline (NA), and compared its effects with those exerted by amphetamine. These studies have been reviewed by Kalix (1990) and below is a summary of the findings.

Using tissues loaded with ^{3}H-DA, cathinone application caused a rapid and reversible increase in the release of radioactivity from isolated rabbit caudate nucleus, rat striatal slices, and rat nucleus accumbens with an effectiveness half of amphetamine and eight times more than cathine. The increase in DA was due to cathinone inhibiting reuptake of this neurotransmitter. Similar results were found with rat neostriatum synaptosomes. For 5-HT, no change in whole brain levels was found (Nielsen 1985; Wagner et al. 1982); however, cathinone caused release of radioactivity from tritiated 5-HT loaded rat caudate nucleus similar to amphetamine (Kalix 1984). Cathinone also caused increased release of radioactivity from isolated rabbit atrium loaded with ^{3}H-NA supporting its sympathomimetic action.

In recent years, more information has become available on the mechanism of action of cathinone as it is often included in the list of synthetic cathinones being studied. As part of their study on behavioral effects of cathinone, mephedrone, and 3,4-methylenedioxymethamphetamine (MDMA), Shortall et al. (2013a) found no significant effect on levels of DA and 5-HT in the rat striatum, frontal cortex, or hippocampus, 7 days after intermittent dosing with cathinone (1 or 4 mg/kg) and 2 h after acute injection of cathinone (4 mg/kg). However, in another study, Shortall et al (2013b) found elevated levels of 5-HT in the rat striatum and hypothalamus 2 h after cathinone injection (4 or 10 mg/kg). Simmler et al. (2013) examined (1) the potency of cathinone to inhibit transporter proteins for DA, NA, and 5-HT (DAT, NET and SERT, respectively) in transfected human embryonic kidney HEK 293 cells, and (2) its binding affinity to monoamine transporters, DA receptors, adrenoceptors, 5-HT receptors, trace-amine associated receptor (TA_1), and histamine receptor. Cathinone inhibited NET, DAT, and SERT with IC_{50} values of 0.199, 14.0, and >100 μM, respectively, showing DAT/SERT inhibition ratio >10, which is correlated with drugs producing psychostimulant effects in humans. It also released DA through DAT. In addition, cathinone bound to α_1-adrenoceptors and showed low affinity for TA_1 receptors. The latter plays an important role in modulation of dopaminergic and serotoninergic activity (Lindemann et al. 2008; Revel et al. 2011). Based on the transporter inhibition profile, Simmler et al. (2013) placed cathinone in "methamphetamine-like cathinones" group, a group that preferentially show catecholamine uptake inhibition and acts as DA releaser. Interestingly, cathinone showed very low blood–brain barrier (BBB) permeability. This low BBB permeability of cathinone is at odds with the view that CNS effectiveness of cathinone is due to its presumed lipophilic property (Kalix 1991) and, hence, crosses the BBB more readily than cathine or norephedrine. However, their finding of very low BBB permeability may explain the lack of gross BBB disruption, observed with Evans Blue and Trypan Blue dyes, in mice treated with fresh khat extract despite showing a hyperthermic response (Gerald and Patel 2017).

3.6 Effects of Khat in Humans

The positive effects of khat on humans are well documented. Its use produces sensation of euphoria and alertness, relief from fatigue, excitement, feeling of energy, and loquacity. Withdrawal effects are mild and range from depression, anxiety, or nightmares. In persons chewing khat, Wondemagegn et al. (2017) found 25% increase in depression and a fivefold increase in anxiety. Long-term use results in adverse health effects on other physiological systems (Table 3.1). In some individuals, excessive use, such binge intake, can result in psychotic symptoms, which usually abate after cessation of the khat use. Khat users also show higher correlation for psychosis than nonusers, though whether khat causes these episodes or is related to stress and mental trauma the individuals endured or is due to underlying susceptibility is not clear [reviewed by (Odenwald 2014; Widmann et al. 2014)].

A number of studies have shown negative cognitive effects in regular khat users. In an aviation examination, Khattab and Amer (1995) found that regular khat users have poorer performance in visual memory, perception, long-term memory, and perceptual speed compared to occasional and non-khat users. Impaired inhibitory

Table 3.1 Some common acute and chronic effects of khat use

Acute effects chronic effects	*After effects (withdrawal) of acute khat use*
Euphoria Alertness Loquacity Relief from fatigue Insomnia Anorexia Hyperthermia Tachycardia Hypertension Mydriasis	Depressive mood Poor sleep quality Nightmares Lethargy Irritability
Oral and dental health	*Gastrointestinal tract*
Periodontal disease Mucosal lesions	Gastritis Enteritis Constipation
Reproductive health	*Cardiovascular*
Impaired sexual function Spermatorrhoea Low birth weight	Arrhythmias Infraction Palpitations
Liver toxicity	*Central Nervous System*
	Psychotic episodes Possible cognitive and executive function impairment

control (Colzato et al. 2011a), deficits in working memory and cognitive flexibility (Colzato et al. 2011b; Hoffman and Al'Absi 2012) have also been reported. Khat being a psychostimulant also causes sleep disturbances (Nakajima et al. 2014), and this may contribute to the emotional state of the person after a khat session.

3.7 Dependence and Addiction to Khat

The 14th Report of the WHO Expert Committee on Dependence (1965) stated: "... with respect to khat, confirming the view that abuse of this substance is a regional problem and may be best controlled at that level." In 1980, WHO classified khat as a "drug of abuse" that produces mild to moderate psychological dependence and did not consider it to be seriously addictive (Al-Mugahed 2008). It was ranked below tobacco and alcohol (Nutt et al. 2007) in terms of its dependence and addictive potential. However, while there may be reservations on the potential of khat to produce dependence and addiction in the user, this is not the case with cathinone, which shows in different methods and tests that it has a reinforcing and addictive potential. Yanagita (1979, 1986) reported that rhesus monkeys self-administer cathinone in sprees of 6–59 h. In comparison to cocaine, L-cathinone maintained a higher rate (Schuster and Johanson 1979) and D,L-cathinone had similar reinforcing effects (Woolverton and Johason 1984). Gosnell et al. (1996)

found that blocking the D1-DA receptor increased self-administration in rats but blocking D2-DA receptor had no effect, an observation suggesting a role of D1-DA receptors in mediating reinforcing effects of cathinone.

Rats given cathinone either by intravenous or intracerebroventricular route showed conditioned place preference (CPP), which was reduced by pretreatment with CGS 10746B, a DA release inhibitor (Calcagnetti and Schechter 1993; Schechter 1991). Recently, CPP was also shown to be induced in mice treated with methanolic fresh khat extract (Kilkwang and Patel 2017).

Until recently, few studies using validated criteria for assessing dependence and abuse potential of drugs of abuse have been carried out in khat users. Widler et al. (1994) gave drug-naïve volunteers khat to chew, and a significant abuse potential was assessed using the Addiction Research Center Inventory (ARCI). Using Severity of Dependence Scale, a significant correlation between psychological dependence and the salivary levels of cathinone, cathine, and norephedrine was found among 30 khat users (Gossop et al. 1995; Kassim et al. 2010, 2012). Male khat chewers exhibited more dependence symptoms compared to females, and females show a positive correlation between age and khat dependence (Nakajima et al. 2014). A prevalence dependence rate of 48.9% was found among khat chewers, with males starting at a younger age (14.8 years) than females (19.5 years) (Nakajima et al. 2017). Similar findings have been reported using the DSM-IV (1994) criteria for substance dependence syndrome, where three out of the seven criteria for dependence were meet in a third of the study subjects: 13% increased khat use, 19% failed cessation attempts, 17% having withdrawal symptoms (depression, increase in appetite, and sleep pattern disruption), and overall 38% reporting continued khat use despite adverse health impact (Kassim et al. 2013). Tolerance to khat use (66.7%), withdrawal symptoms (94.0%), failure to stop or reduce (78.8%), continued use despite adverse health outcomes (93.3%), more use than wished (72.2%), and impact on social life (97.0%) was also found in a sample of Somalis living in Kenya (Widmann et al. 2014). These studies revealed that khat can produce dependence and addiction at a higher rate than previously presumed and further studies are warranted.

3.8 Conclusions

The fresh leaves and twigs of *C. edulis* Forsk., khat, is chewed by millions of persons in the parts of the Arabian Peninsula and along Eastern Africa—a practice with deep roots in their culture and tradition. Its active agent, cathinone, is an intermediate metabolite in the biosynthetic pathways in both *C. edulis* and *Ephedra* spp. It is a psychostimulant, and like other psychostimulants, such as amphetamine and cocaine, inhibits DAT and NET, and like amphetamines stimulates release of DA. In countries where there is extensive commercial cultivation such as Ethiopia, Yemen, and Kenya, it is an important cash crop. Recent studies show that khat use results in dependence and addiction as well as producing adverse somatic health effects. However, it is not likely that in the khat using countries its use will be restricted or made illegal and, therefore, future understanding of its short- and long-term effects in the users are needed to educate the user and for policy decisions on reducing its use.

References

Al-Motarreb A, Baker K, Broadley KJ (2002) Khat: pharmacological and medical aspects and its social use in Yemen. Phytother Res 16:403–413. https://doi.org/10.1002/ptr.1106

Al-Mugahed L (2008) Khat chewing in Yemen: turning over a new leaf: khat chewing is on the rise in Yemen, raising concerns about the health and social consequences. Bull World Health Organ 86(10):741–742

Alles GA, Fairchild MD, Jensen M (1961) Chemical pharmacology of *Catha edulis*. J Med Pharm Chem 3:323–352

Anderson D, Beckerleg S, Hailu D, Klein A (eds) (2007) The Khat controversy: stimulating the debate on drugs. Berg, Oxford

Appendino G, Minassi A, Taglialatela-Scafati O (2014) Recreational drug discovery: natural products as lead structures for synthesis of smart drugs. Nat Prod Rep 31(7):880–904

Atlabachew M, Combrinck S, Viljoen AM et al (2016) Isolation and in vitro permeation of phenylpropylamino alkaloids from Khat (*Catha edulis*) across oral and intestinal mucosal tissues. J Ethnopharmacol 194:307–315. https://doi.org/10.1016/j.jep.2016.09.012

Bailey WJ (1995) Methcathinone. Indiana Prevention Resource Center. https://www.nurseslearning.com/courses/corexcel/cxnrp-1600/chap6/course/chap1/P9.html. Accessed 13 Nov 2017

Beitter A (1901) Pharmakognostich-chemische Untersuchung der *Catha edulis*. Arch Pharm (Weinheim) 239:17–33

Calcagnetti DJ, Schechter MD (1993) Place preference for the psychostimulant cathinone is blocked by pretreatment with a dopamine release inhibitor. Prog Neuropsychopharmacol Biol Psychiatry 17:637–649

Colzato LS, Ruiz MJ, van den Wildenberg WP et al (2011a) Long-term effects of chronic khat use: impaired inhibitory control. Front Psychol 1:219. https://doi.org/10.3389/fpsyg.2010.00219

Colzato LS, Ruiz MJ, van den Wildenberg WP, Hommel B (2011b) Khat use is associated with impaired working memory and cognitive flexibility. PLoS ONE 6:e20602. https://doi.org/10.1371/journal.pone.0020602

Cox G, Rampes H (2003) Adverse effects of khat: a review. Adv Psych Treat 9:456–463

Emerson TS, Cisek JE (1993) Methcathinone: a Russian designer amphetamine infiltrates the rural Midwest. Ann Emerg Med 22(12):1897–1903

en.wikipedia.org/wiki/khat. Accessed 13 October 2017

Feyissa AM, Kelly JP (2008) A review of the neuropharmacological properties of khat. Progress Neuropsychopharmacol Biol Psych 32(5):1147–1166. https://doi.org/10.1016/j.pnpbp.2007.12.033

Gebissa E (2004) Leaf of Allah: khat & agriculuture transformation in Harerge, Ethiopia 1875–1991. Ohio University Press, Athens, Ohio

Geisshusler S, Brenneisen R (1987) The content of psychoactive phenylpropyl and phenylpentenyl khatamins in *Catha edulis* Forsk of different origins. J Ethnopharmacol 19:269–277

Gerald O, Patel NB (2017) Lack of gross impairment of blood brain barrier integrity with binge-dose pattern of khat (*Catha edulis* Forsk) extract treatment. Society of Neuroscientists of Africa conference Abs No. MP-08 11–14 June 2017, Entebbe, Uganda

Glennon RA, Young R, Martin BR, Dal Cason TA (1995) Methcathinone ("cat"): an enantiomeric potency comparison. Pharmacol Biochem Behav 50(4):601–606

Goerig M, Bacon D, van Zundert A (2012) Carl Koller, cocaine, and local anesthesia: some less known and forgotten facts. Reg Anesth Pain Med 37(3):318–324. https://doi.org/10.1097/aap.0b013e31825051f3

Gosnell BA, Yracheta JM, Bell SM, Lane KE (1996) Intravenous self-administration of cathinone by rats. Behav Pharmacol 7(6):526–531

Gossop M, Darke S, Griffiths P et al (1995) The severity of dependence scale (SDS): psychometric properties of the SDS in English and Australian samples of heroin, cocaine and amphetamine users. Addiction 90:607–614

Groves RA, Hagel JM, Zhang Y et al (2015) Transcriptome profiling of Khat (*Catha edulis*) and *Ephedra sinica* reveals gene candidates potentially involved in amphetamine-type alkaloid biosynthesis. PLoS ONE 10(3):e0119701. https://doi.org/10.1371/journal.pone.0119701

Grue-Sorenson G, Spenser ID (1994) Biosynthetic route to the Ephedra alkaloids. J Am Chem Soc 166:6195–6200

Guantai A, Maitai C (1982) Relative distribution of cathinone and norpseudoephedrine in *Catha edulis* growing in Kenya. EAMJ 59:394–398

Hagel JM, Krizevski R, Kilpatrick K et al (2011) Expressed sequence tag analysis of khat (*Catha edulis*) provides a putative molecular biochemical basis for the biosynthesis of phenylpropylamino alkaloids. Genet Mol Biol 34(4):640–646. https://doi.org/10.1590/s1415-47572011000400017

Hersi A, Abdalla M (2013) Sharia law and legality of consumption of khat (*Catha edulis*): views of Australian Imams. Int J Humanit Soc Sci 3(21):248–257

Hoffman R, Al'Absi M (2012) Working memory and speed of information processing in chronic khat users: preliminary findings. Eur Addict Res 19:1–6

Hunter FM (1877) An account of the British Settlement of Aden in Arabia. Frank Cass and Company Ltd Publishers, 1st edn. New impression 1968

Hyde JF, Browning E, Adams R (1928) Synthetic homologs of d, l-ephedrine. J Am Chem Soc 50 (8):2287–2292

Kalix P (1984) Effect of the alkaloid (-)-cathinone on the release of radioactivity from rat striatal tissue prelabelled with ^{3}H-serotonin. Neuropsychobiology 12(2–3):127–129

Kalix P (1990) Pharmacological properties of the stimulant khat. Pharmac Ther 48:397–416

Kalix P (1991) The pharmacology of psychoactive alkaloids from ephedra and catha. J Ethnopharmacol 32:201–208

Kassim S, Islam S, Croucher R (2010) Validity and reliability of a Severity of Dependence Scale for khat (SDS-khat). J Ethnopharmacol 132(3):570–577. https://doi.org/10.1016/j.jep.2010.09.009

Kassim S, Hawash A, Johnston A, Croucher R (2012) Validation of self-reported khat chewing amongst khat chewers: an exploratory study. J Ethnopharmacol 140(1):193–196. https://doi.org/10.1016/j.jep.2011.12.040

Kassim S, Croucher R, Al'Absi M (2013) Khat dependence syndrome: a cross sectional preliminary evaluation amongst UK-resident Yemeni khat chewers. J Ethnopharmacol 146 (3):835–841. https://doi.org/10.1016/j.jep.2013.02.011

Khattab NY, Amer G (1995) Undetected neuropsychological sequelae of khat chewing in standard aviation medical examination. Aviat Space Environ Med 66(8):739–744

Kilkwang L, Patel NB (2017) *Catha edulis* Forsk (Khat) induces conditioned place preference in mice. Society of Neuroscientists of Africa Abs no MP-07. 11–14 June 2017, Entebbe, Uganda

Klein A, Beckerleg S (2007) Building castles of spit—the role of khat chewing in worship, work and leisure. In: Goodman J, Lovejoy P, Sherrat A (eds) Consuming habits. Routledge, London, pp 238–254

Krizevski R, Dudai N, Bar E, Lewinsohn E (2007) Developmental patterns of phenylpropylamino alkaloids accumulation in khat (*Catha edulis*, Forsk.). J Ethnopharmacol 114(3):432–438. https://doi.org/10.1016/j.jep.2007.08.042

Lee MR (2011) The history of ephedra (ma-huang). J R Coll Physicians Edinb 41:78–84

Lindemann L, Meyer CA, Jeanneau K et al (2008) Trace amine-associated receptor 1 modulates dopaminergic activity. J Pharmacol Exp Ther 324(3):948–956. https://doi.org/10.1124/jpet.107.132647

Magdum SS (2011) An overview of khat. Addict Disord Treat 10(2):72–83. https://doi.org/10.1097/adt.0b013e3181f002db

Nabuzoka D, Badhadhe FA (2000) Use and perception of khat among young Somalis in a UK city. Addict Res 8:5–26

Nakajima M, Dokam A, Alsameai A et al (2014) Severity of khat dependence among adult khat chewers: the moderating influence of gender and age. J Ethnopharmacol 155(3):1467–1472. https://doi.org/10.1016/j.jep.2014.07.030

Nakajima M, Hoffman R, Al'Absi M (2017) Level of khat dependence, use patterns, and psychological correlates in Yemen: a cross-sectional investigation. EMHJ 23(3):161–167

Nielsen J (1985) Cathinone affects dopamine and 5-hydroxytryptamine neurons in vivo as measured by changes in metabolites and synthesis in four forebrain regions in the rat. Neuropharmacology 24:845–852

Nutt D, King LA, Saulsbury C, Blakemore C (2007) Development of a rational scale to assess the harm of drugs of potential misuse. Lancet 369(9566):1047–1053. https://doi.org/10.1016/s0140-6736(07)60464-4

Odenwald M (2014) Mental health problems associated with the use and abuse of khat (*Catha edulis*). In: Bentivoglio M, Cavalheiro EA, Kristensson K, Patel NB (eds) Neglected tropical diseases and conditions of the nervous system. Springer, New York, pp 293–305. https://doi.org/10.1007/978-1-4614-8100-3_16

Odenwald M, Lingenfelder B, Schauer M et al (2007) Screening for posttraumatic stress disorder among Somali ex-combatants: a validation study. Confl Health 1:10. https://doi.org/10.1186/1752-1505-1-10

Parke D (1957) Ethylaminopropiophenone compounds. US Patent 2802865, 13 August 1957

Patel NB (2015) "Natural Amphetamine" khat: a cultural tradition or a drug of abuse? In: Taba P, Lees A, Sikk K (eds) International review of neurobiology, vol 120. Academic Press, Burlington, pp 235–255

Patel SL, Wright S, Gammampila A (2005) Khat use among Somalis in four English cities. Home Office Online Report 47/05, United Kingdom. Accessed 13 Nov 2017

Revel FG, Moreau JL, Gainetdinov RR et al (2011) TAAR1 activation modulates monoaminergic neurotransmission, preventing hyperdopaminergic and hypoglutamatergic activity. Proc Natl Acad Sci USA 108(20):8485–8490. https://doi.org/10.1073/pnas.1103029108

Saha S, Dollery C (2006) Severe ischaemic cardiomyopathy associated with khat chewing. J R Soc Med 99(6):316–318. https://doi.org/10.1258/jrsm.99.6.316

Schechter MD (1991) Effect of learned behavior upon conditioned place preference to cathinone. Pharmacol Biochem Behav 38(1):7–11

Schorno X, Steinegger E (1978) The phenylalkylamines of *Catha edulis* Forsk.: the absolute configuration of cathinone. United Nations Document (MNAR/7/1978)

Schorno X, Brennneisen R, Steinegger E (1982) Qualitative und quantative Untersuchungen uber das Vorkommen ZNS-aktiver phenylpropylamine in Handelsdrogen and uber deren Verteilung in verschiedenen organen von *Catha edulis*. Pharm Acta Helva 57:168–176

Schuster CR, Johanson CE (1979) Behavioral studies of cathinone in monkeys and rats. In: Problems of drug dependence. NIDA Research Monograph 27, Proceedings of 41st annual scientific meeting of the committee on problems of drug dependence. US Government Printing Office, Washington, pp 324–325

Shortall SE, Macerola AE, Rabbi TR et al (2013a) Behavioural and neurochemical comparison of chronic intermittent cathinone, mephedrone and MDMA administration to the rat. Eur Neuropsychopharmacol 23(9):1085–1095. https://doi.org/10.1016/j.euroneuro.2012.09.005

Shortall SE, Green AR, Swift KM et al (2013b) Differential effects of cathinone compounds and MDMA on body temperature in the rat, and pharmacological characterization of mephedrone-induced hypothermia. Br J Pharmacol 168(4):966–977. https://doi.org/10.1111/j.1476-5381.2012.02236.x

Simmler LD, Buser TA, Donzelli M et al (2013) Pharmacological characterization of designer cathinones in vitro. Br J Pharmacol 168(2):458–470. https://doi.org/10.1111/j.1476-5381.2012.02145.x

Szendrei K (1980) The chemistry of khat. Bull Narc 32(3):5–35

Toennes SW, Harder S, Schramm M et al (2003) Pharmacokinetics of cathinone, cathine and norephedrine after chewing khat leaves. Br J Clinical Pharmacol 56(1):125–130

United Nations Office on Drugs and Crime (2015) World Drug Report, p 1

von Brücke T (1941) Über die zentral erregende Wirkung des Alkaloides Cathin. Naunyn Schmiedeberg's Archiv für experimentelle Pathologie und Pharmakologie 198:100

Wagner G, Preston K, Ricaurte G et al (1982) Neurochemical similarities between d, l-cathinone and d-amphetamine. Drug Alcohol Depend 9:279–284

Ward C, Gatter P (2000) Qat in Yemen—towards a policy and action plan. Document of the World Bank, 108 pp. http://www.yemenwater.org/wp-content/uploads/2013/03/ward-gatter-2000-qat-in-yemen-pdf.pdf. Accessed 13 Nov 2017

Widler P, Mathys K, Brenneisen R et al (1994) Pharmacodynamics and pharmacokinetics of khat: a controlled study. Clin Pharmacol Ther 55(5):556–562

Widmann M, Warsame AH, Mikulica J et al (2014) Khat use, PTSD and psychotic symptoms among Somali refugees in Nairobi—a pilot study. Front Public Health 2:71. https://doi.org/10.3389/fpubh.2014.00071

Wolfes O (1930) Uber das Vorkommen von D-nor-iso-Ephedrin in *Catha edulis*. Arch Pharm (Weinheim) 268:81–83

Wondemagegn AT, Cheme MC, Kibret KT (2017) Perceived psychological, economic, and social impact of khat chewing among adolescents and adults in Nekemte Town, East Welega Zone, West Ethiopia. Biomed Res Int. https://doi.org/10.1155/2017/7427892

Woolverton WL, Johason CE (1984) Preference in rhesus monkeys given a choice between cocaine and d, l-cathinone. J Exp Anal Behav 41(1):35–43

World Health Organization Expert Committee on Dependence Producing Drugs (1965) 14th Report. Technical Report Series No. 312

World Health Organization (2008) Khat chewing in Yemen: turning over a new leaf. Bull World Health Organ 86(10):741–742

Yanagita T (1979) Studies on cathinones: cardiovascular and behavioural effects in rats and self-administration experiments. In: Problems of drug dependence. NIDA Research Monograph 27, Proceedings of 41st Annual scientific meeting of the committee on problems of drug dependence. US Government Printing Office, Washington, pp 326–327

Yanagita T (1986) Intravenous self-administration of cathinone and 2-amino-l-(2,5-dimethoxy-4-methyl)-phenylpropane in rhesus monkeys. Drug Alcohol Depend 17(2–3):135–141

Zahran MA, Khedr A, Dahmash A, El-Ameir YA (2014) Qat farms in Yemen: ecology, dangerous impacts and future promise. Egypt J Basic Appl Sci I:1–8. https://doi.org/10.1016/j.ejbas.2013.09.002

Chapter 4
Analytical Methods Used for Identification and Determination of Synthetic Cathinones and Their Metabolites

Dariusz Zuba and Piotr Adamowicz

Abstract Synthetic cathinones are a broad group of chemicals, which have very similar structure. Many substances seized from the drug market are isomers, homologues, analogues, etc. They follow similar metabolism processes, and often convert in the human body to the same metabolites; some being also cathinone derivatives. Moreover, active doses differ significantly among drugs, causing that the levels of cathinones and their metabolites in the body fluids and tissues are in a broad range, from low ng/mL to hundreds mg/L. A variety of analytical techniques is applied to identify synthetic cathinones in seized drugs. Unequivocal identification of an active substance is crucial, especially in countries with individual drug control system, as the legal consequences for, e.g. possession of different isomers may vary substantially. Gas chromatography–mass spectrometry (GC-MS) is the most commonly method used for preliminary identification in forensic laboratories, but it has to be supported by other techniques, e.g. by Fourier-transformed infrared spectrometry (FTIR), nuclear magnetic resonance (NMR) and liquid chromatography with different detectors, mainly tandem mass spectrometers. Synthetic cathinones are analysed in different biological matrices, including blood, serum, plasma, dried blood spots, urine, hair, oral fluid and postmortem body tissues. Samples are prepared for the analysis, e.g. by dilution, precipitation, liquid–liquid extraction (LLE) and solid phase extraction (SPE). Enzyme hydrolysis (especially for urine), washing out or digestion (for hair), and derivatization are also included in some procedures. Triple quadrupole LC-MS/MS systems are the most frequently used. Many analytical challenges cause that more sophisticated techniques, including liquid chromatography-high-resolution mass spectrometry (LC-HRMS), are increasingly applied in the analysis of biological materials for the identification and quantitation of cathinones.

D. Zuba (✉) · P. Adamowicz
Institute of Forensic Research, Westerplatte 9, 31-033 Kraków, Poland
e-mail: dzuba@ies.krakow.pl

P. Adamowicz
e-mail: padamowicz@ies.krakow.pl

J. B. Zawilska (ed.), *Synthetic Cathinones*, Current Topics in Neurotoxicity 12,
https://doi.org/10.1007/978-3-319-78707-7_4

Keywords Synthetic cathinones · Seized material · Biological material Analytical methods · GC-MS · LC-MS · HPLC-DAD

Abbreviations

ATR	Attenuated total reflectance technique
CE	Capillary electrophoresis
CI	Chemical ionization
DART-MS	Direct analysis in real time-mass spectrometry
DAD	Diode array detection
DBS	Dried blood spots
DESI-MS	Desorption electrospray ionization-mass spectrometry
DLLME	Dispersive liquid–liquid microextraction
EI	Electron impact ionization
ELISA	Enzyme-linked immunosorbent assay
ENFSI	European Network of Forensic Science Institutes
ESI	Electrospray ionization
FTIR	Fourier-transformed infrared spectrometry
GC	Gas chromatography
GC-EI-MS	Gas chromatography–electron impact–mass spectrometry
GC-FID	Gas chromatography with flame ionization detector
GC-MS	Gas chromatography–mass spectrometry
HPLC-DAD	High-performance liquid chromatography with diode array detection
HRMS	High-resolution mass spectrometry
IMS	Ion mobility spectrometry
IR	Infrared spectroscopy
LC	Liquid chromatography
LC-HRMS	Liquid chromatography-high resolution mass spectrometry
LC-HRMS/MS	Liquid chromatography-high resolution multiple mass spectrometry
LC-MS	Liquid chromatography–mass spectrometry
LC-MS/MS	Liquid chromatography-tandem mass spectrometry
LC-ESI-QTOFMS	Liquid chromatography-electrospray ionization-quadrupole time-of-flight mass spectrometry
LC-QTOFMS	Liquid chromatography-quadrupole time-of-flight mass spectrometry
LLE	Liquid–liquid extraction
LOD	Limit of detection
MALDI-TOF	Matrix-assisted laser desorption/ionization time-of-flight mass spectrometry
MRM	Multiple reaction monitoring
MS	Mass spectrometry
MS/MS	Tandem mass spectrometry
NIR	Near infrared

NMR	Nuclear magnetic resonance
μ-SPE	Miniaturized solid phase extraction
PP	Protein precipitation
QqQ	Triple quadrupole
QTOF	Hybrid quadrupole time-of-flight analyser
SALLE	Salting out liquid–liquid extraction
SIM	Selected ion monitoring
SLE	Supported liquid extraction
SPE	Solid phase extraction
SPME	Solid phase microextraction
SWGDRUG	Scientific Working Group for the Analysis of Seized Drugs
RF-MS-MS	RapidFire tandem mass spectrometry
UHPLC-DAD	Ultra high-performance liquid chromatography with diode array detection
UHPLC-MS	Ultra high-performance liquid chromatography–mass spectrometry
UV/VIS	Ultraviolet/visible spectroscopy
VAMS	Volumetric absorptive microsampling

4.1 Introduction

Rapid increase in the number of substances offered to potential users and their very similar chemical structures caused that in recent years analytical protocols used for examination of narcotic drugs and psychotropic substances seized from the drug market and in biological material have significantly changed. The analytical methods, which were routinely used in forensic laboratories for many years, such as gas chromatography with flame ionization detector (GC-FID) or high-performance liquid chromatography with diode array detector (HPLC-DAD), turned out to be insufficient due to their limited selectivity. Even gas chromatography–mass spectrometry (GC-MS) has the limited application to the investigation of new psychoactive substances.

Modern analytical techniques expand research capabilities, but when used alone they may give false results. For this reason, a comprehensive approach to identify new substances from different chemical groups has to be applied. Because in many cases the seized preparations are mixtures, the hyphenated techniques are the best choice. Separation of the analytes is usually achieved by gas chromatography (GC) or liquid chromatography (LC), while their characteristics are determined by mass spectrometry (MS). The gas chromatographs are usually connected to a quadrupole mass detector, while a liquid chromatograph to triple quadrupole (QqQ) or a hybrid quadrupole time-of-flight analyser (QTOF). In the analysis of seized drugs, Fourier transform infrared spectrometry (FTIR) and nuclear magnetic resonance spectroscopy (NMR) are commonly used for structure elucidation, especially when pure substances are subject for investigation. When selecting a new technique for the analytical procedure of identification of cathinones, its complementarity to the previously used is the main criteria. The idea is that the new technique improves knowledge on the structure of the investigated compounds and eliminates the potential interferences from the matrix.

Analytical screening methods used in forensic and clinical toxicology laboratories to reveal the presence of synthetic cathinones in biological materials should be rapid, sensitive and specific. Unfortunately, there is a lack of comprehensive analytical screening methods available for simultaneous detection of such novel drugs of abuse. Existing methods are not devoid of defects, and the continually emerging new drugs on the illicit drug market makes that they also must be continually modified.

Synthetic cathinones were analysed in variety of biological matrices including blood, serum, plasma, dried blood spots, urine, dried urine, hair, oral fluid and postmortem body tissues. The methods covered parent compounds but also metabolites were included. The most common sample preparations were dilution, precipitation (PP), liquid–liquid extraction (LLE) and solid phase extraction (SPE). Analytical methods utilizing immunoassay, GC-MS, HPLC-DAD, liquid chromatography–mass spectrometry (LC-MS), liquid chromatography-tandem mass spectrometry (LC-MS/MS), liquid chromatography-high-resolution mass spectrometry (LC-HRMS) and direct analysis in real time-mass spectrometry (DART-MS) were reported in analysis of biological materials. In this chapter, analytical procedures that allow unambiguous identification and quantitation of synthetic cathinones in samples seized from the drug market and in biological material are presented.

4.2 Analysis of Seized Materials

4.2.1 Presumptive Tests

Colour tests are an important tool used for screening for licit and illicit drugs, in spite of developments in instrumental technology and the increased portability of analytical instruments which are more and more widely used in the field. The popularity of colour tests arises from the fact that they are generally simple, rapid, inexpensive and relatively sensitive and selective. Colour tests can be used also by people without extensive chemical backgrounds. However, synthetic cathinones are a class of psychoactive substances that do not currently have a suitable, specific colour test regularly used for presumptive identification. It was shown that methcathinone and its analogues including *N*,*N*-dimethylmethcathinone (*N*,*N*-DMC) and 4-methylmethcathinone (4-MMC; mephedrone) produced bright yellow colours in reaction with both Marquis reagent and Liebermann's reagent, while other colour test failed to give coloured products with these drugs (Toole et al. 2012). In 2016, a report was published on a presumptive colour test method employing reagent copper(II)-2,9-dimethyl-1,10-phenanthroline (Cu(II)-neocuproine) for screening of unknown samples for cathinones. The procedure was based on the reduction of copper(II) to copper(I) in the presence of the drug, followed by formation of a yellow-orange coloured complex with a maximum absorbance at 453 nm. A false negative response rate was slightly greater than 10%. Other recreational drugs, adulterants and white powders tested afforded a yellow colour change in 10 out of 83 substances examined; however, most of these were still distinguishable from a true positive test result (Philp et al. 2016).

Ion mobility spectrometry (IMS) has been considered as an analytical technique of choice to detect illicit substances in security points in airports, borderlines and customs. Databases of the reduced mobility (K_0) values of illicit drugs are available in the scientific literature and they are updated with data for emerging drugs. Joshi and coworkers evaluated the performance of the standard ^{63}Ni ionization IMS technique for the screening and identification of cathinones and associated psychoactive compounds. The results showed that the IMS is an acceptable rapid and efficient screening tool for cathinones, positively detecting at least one cathinone in 77% of the samples tested (Joshi et al. 2014). In another paper, the authors evaluated the influence of different measurement conditions on the reading, and determined K_0 values of different families of new psychoactive substances (NPS), including phenethylamines, cathinones, synthetic cannabinoids and tryptamines (Armenta et al. 2015).

Desorption electrospray ionization-mass spectrometry (DESI-MS) is a technique that allows for the analysis of compounds directly from ambient surfaces, reducing or eliminating the need for sample preparation. The application of the DESI-MS technique to the fast presumptive detection and chemical analysis of mephedrone was presented (Stojanovska et al. 2014). A particularly suitable surface, semi-porous polytetrafluoroethylene (PTFE, Teflon) was utilized, as it generated the least variable signal and high reproducibility compared to other tested surfaces.

Another ambient ionization method, i.e. a direct analysis in real time-mass spectrometry (DART-MS), was employed to rapidly screen cathinones, alone and in mixtures, readily enabling differentiation of the active drug(s) from various cutting agents. Accurate mass determinations provided preliminary identification of the various components of drug mixtures (Musah et al. 2014).

4.2.2 *Preliminary Identification by Gas Chromatography–Mass Spectrometry Method*

The GC-MS method is a basic investigative tool used in drug forensic laboratories. This is due to the fact that it can obtain reproducible mass spectra, and thus it is possible to create databases, i.e. mass spectral libraries. Commercially available libraries contain ca. 200,000 compounds (e.g. NIST library). Specific databases, e.g. 'Mass spectral database of designer drugs', were also developed (Rosner 2017). However, a common problem with spectra of new substances, including synthetic cathinones, is that since these chemicals are new, they are not included in the libraries. What is more, because of the great similarity in chemical structure of the new compounds to the controlled substances or previously identified NPS, there is a high risk of incorrect (false positive) identification, especially by an inexperienced analyst. Therefore, in recent years closer cooperation between experts in drug analysis has emerged, and systems of sharing analytical data were formed. Noncommercial mass spectra databases are available through Internet websites of the Drugs Working Group of the European Network of Forensic Science Institutes (ENFSI) and Scientific Working Group for the Analysis of Seized Drugs (SWGDRUG).

Synthetic cathinones are derivatives of the parent compound, cathinone, which is one of the components of the psychoactive plant *Catha edulis* Forsk. The chemical structure of cathinone is similar to that of amphetamine; the only difference is the presence of the carbonyl group in the β position of the side chain. Both substances (and their derivatives) contain the basic structure of phenylethylamine (PEA), which is also present in many stimulants and hallucinogenic substances, including those from the so-called 2C, DO_x and NBOMe families. What is more, other popular group of hallucinogens, that is tryptamines, also displays similarities in chemical structures. The comparison of the structures of phenylethylamine, amphetamine, cathinone and tryptamine is shown in Fig. 4.1.

Zuba (2012) described the fragmentation mechanisms of substances from different chemical classes of NPS, including phenethylamines, cathinones and tryptamines; this helped to determine characteristic masses (more precisely, the mass to charge ratios, *m/z*) used for identification. The proposed scheme of preliminary classification of a new psychoactive substance on the basis of (GC-)EI-MS spectrum, presented in Fig. 4.2, is widely used in many laboratories investigating products seized from the drug market.

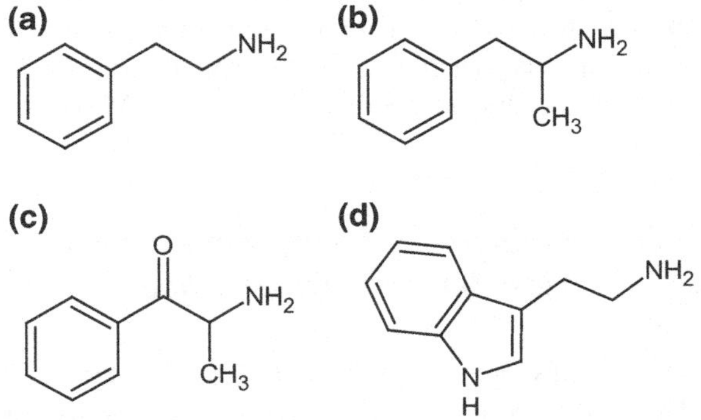

Fig. 4.1 The chemical structure of: **a** phenylethylamine, **b** amphetamine, **c** cathinone and **d** tryptamine

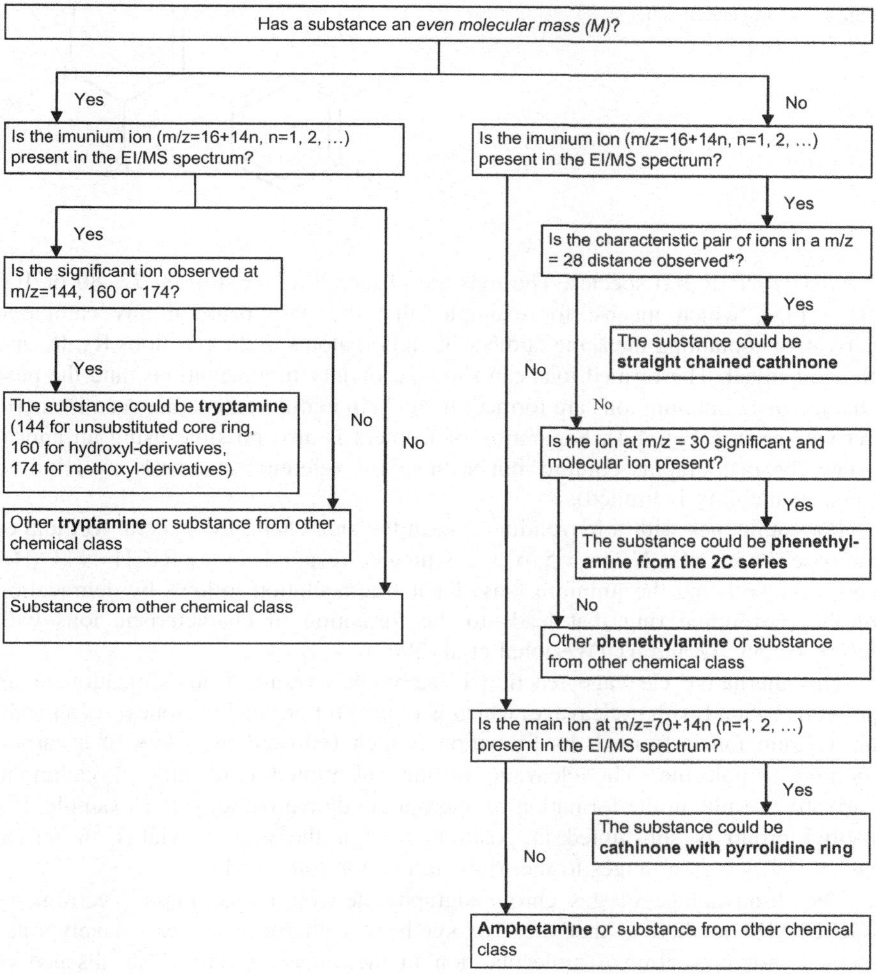

Fig. 4.2 Scheme of preliminary classification of a new psychoactive substance on the basis of EI-MS spectrum

The base peaks in the EI-MS spectra of synthetic cathinones are formed by dissociation of the bond between α and β carbon atoms in the side chain (the general scheme of derivatives of cathinone is presented in Fig. 4.3). As a result of this process, the iminium ion is formed. The mass of the base peak depends on the number of carbon atoms substituted at the positions R_3, R_4 and R_5. The side chain in cathinones is usually aliphatic (and contains a terminal amine group) or it contains a pyrrolidine ring.

In cases of a straight-chained aliphatic side ring, the structural formula of the iminium ion is $C_nH_{2n+2}N^+$ ($n = 1, 2, \ldots$), normally assumed to be represented by

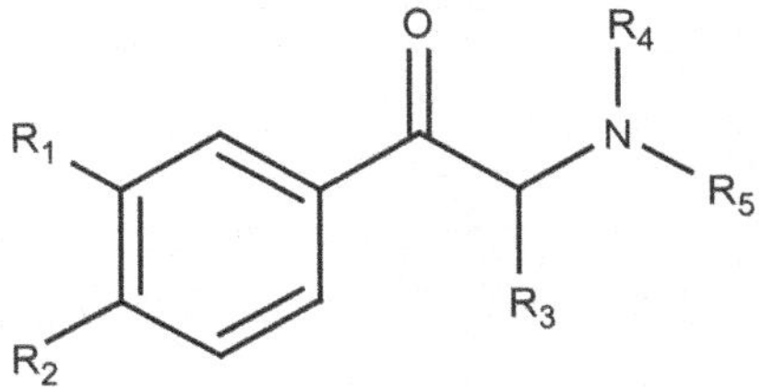

Fig. 4.3 The general scheme of cathinone derivatives

the $R_3CH{=}N^+(R_4R_5)$ species. The ions are observed at *m/z* 44, 58, 72, 86 or 100 (16 + 14*n*), which means, for example, that the base peak of any cathinone derivative containing the same number of carbon atoms in the positions R_3, R_4 and R_5 is identical. The formed ions can show secondary fragmentations, and the next characteristic iminium ions are formed. It can help to differentiate between isomeric derivatives (see Fig. 4.4). Separation of isomers is also possible using an appropriate chromatographic column, but in this case, reference standards are required (their availability is limited).

For cathinones with a pyrrolidine ring in the side chain, the general formula of the base ion is $C_nH_{2n}N^+$ ($n = 5, 6, \ldots$), which corresponds to the $R_3CH{=}N^+(C_4H_8)$ species, also being the iminium ions. Their fragmentation follows by degradation of the pyrrolidine ring that leads to the formation of characteristic ions with *m/z* = 70, 55, 42 and 41 (Westphal et al. 2009).

The alternative cleavage reaction in cathinones results from ionization of an aromatic π-bond with α-cleavage, which is typical for aromatic ketones. As a result, an acylium ion is formed. Its formation is then followed by a loss of a carbon monoxide molecule. The cleavage of the substituted core ring of cathinone derivative results in the formation of appropriate derivative ions. For example, if a methyl group is substituted at position 4 (R_2), the process yields an ion at *m/z* = 107, which arranges to the tropylium cation (*m/z* = 91).

The disadvantage of gas chromatography-electron impact-mass spectrometry (GC-EI-MS) method is that most of synthetic cathinones are extensively fragmented, not presenting a molecular ion in their mass spectra. The absence of molecular ion and/or similar fragmentation pattern among the derivatives may cause the equivocal identification of unknown seized substances. Alternative approach is to use gas chromatography-triple quadrupole-tandem mass spectrometry, and to apply chemical ionization (CI). The implementation of CI source facilitates the determination of molecular mass and thus identification. Developed multiple reaction monitoring (MRM) mode may increase sensitivity and selectivity, which was discussed in Sect. 4.3.2.3. The CI mass spectra and MRM mass spectra of cathinone derivatives can be used as a potential supplemental database along with EI mass spectral database (Gwak et al. 2015).

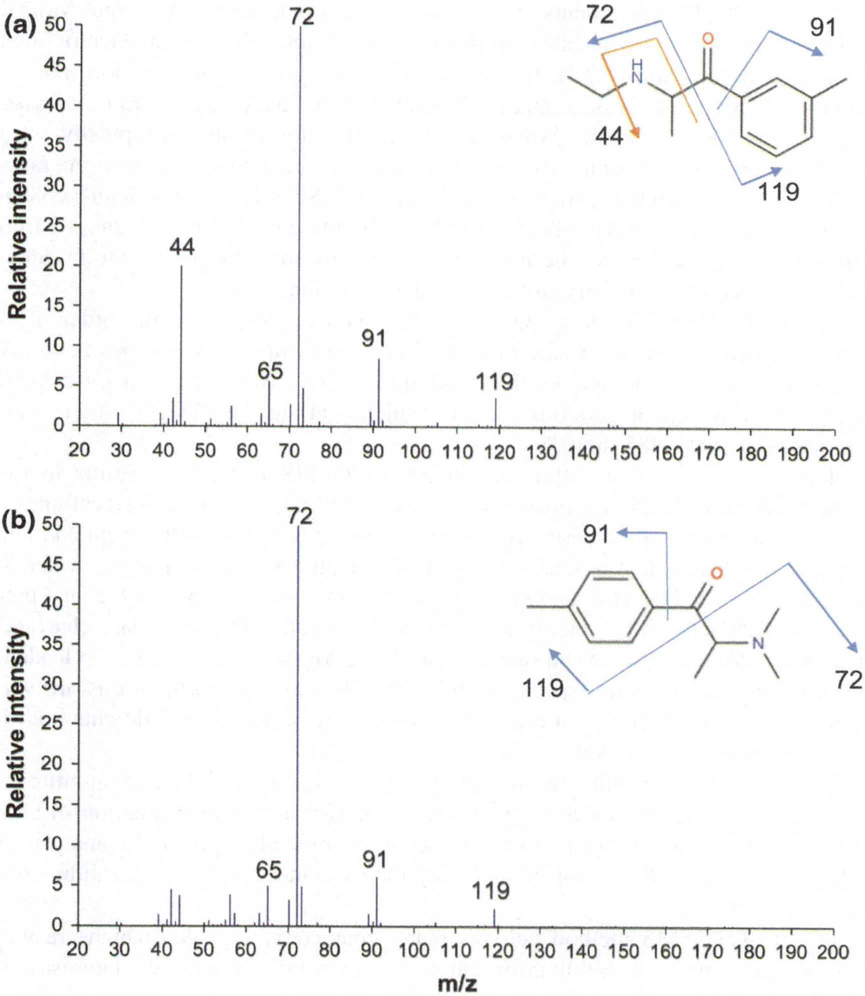

Fig. 4.4 The EI-MS spectra of: **a** 3-MEC and **b** 4-MDMC

4.2.3 Confirmatory Analysis by Liquid Chromatography-Quadrupole Time-of-Flight Mass Spectrometry Method

High-resolution mass spectrometry plays a key role in the structure elucidation and identification of fragmentation patterns of the compounds. For synthetic cathinones, liquid chromatography-quadrupole time-of-flight mass spectrometry (LC-QTOFMS) technique has been used relatively widely (e.g. Nie et al. 2016; Liu et al. 2017; Qian et al. 2017).

The QTOFMS instruments can perform measurements in MS and MS/MS modes. Substances are usually ionized by being sprayed into an electric field (electrospray ionization, ESI). In the analysis of cathinones, positive ion mode is commonly used, and mass spectra are collected in the mass range from low masses (e.g. 60) to 500 *m/z*. In the MS mode, the fragmentor voltage is typically set at 100 V (in order to determine the molecular mass of the compound) or in the range of 200–300 V (to study the main ions). In the MS/MS mode, the fragmentor voltage is also usually in the range of 200–300 V, while the collision energy ranged from 10 to 30 eV. In this mode, the initially formed ions are subjected to the fragmentation process, which is very useful in structure elucidation.

The QTOFMS instruments determine the mass accurately, in the order of 2–5 ppm, depending on the mode of operation. The synthetic cathinones have low molecular masses, usually less than 250 amu. In this case, an accuracy error of about 2 ppm corresponds to 0.0005 amu, which is at the same level as the mass of an electron (0.0005485 amu).

The fragmentation of cathinones in ESI-QTOFMS instrument results in two most commonly observed transitions. The spectra of the straight-chained cathinones are characterized by an intense ion at *m/z* value lower than 18 when compared to the pseudomolecular ion. It indicates the loss of a neutral molecule, namely water. In the case of ESI-QTOFMS, the secondary fragmentations are also possible, and their efficiency depends on the energy delivered to the species. The secondary cleavages take place mainly on α-carbon (adjacent to the nitrogen atom), which also leads to the dissociation of neutral species (amine and alkene). The formed ions are very characteristic, and therefore are useful in the determination of constituents both in the core ring and in the side chain.

Fragmentation of cathinones also process by dissociation of the carbon–nitrogen bond. This process produces the ions that are useful in the determination of alkyl substituents to the nitrogen atom, and is predominantly applied to analysis of cathinones being tertiary amines, including the substances with the pyrrolidine ring in the side chain.

LC-ESI-QTOFMS method delivers many characteristic peaks, which are very unique and constitute useful information for investigators who are interested in identifying the structures of new, unknown cathinones.

4.2.4 Other Analytical Methods Used in Identification of Cathinones

HPLC-DAD is an analytical method used in the analysis of seized drugs, including cathinones, for three reasons:

1. Confirmation of a structure of an identified substance;
2. Determination of a location of a substituent in the phenyl ring;
3. Quantitative analysis.

HPLC-DAD instruments allow to obtain the spectra of examined substances in the range of 200–400 nm, after prior separation of the components on a chromatographic column. The selectivity of UV/VIS spectra is limited, therefore these spectra cannot be used for unequivocal identification, but to support and confirm the information obtained from, e.g. mass spectra.

The main advantage of HPLC-DAD method in identification of synthetic cathinones is that the shapes of UV/VIS spectra, including location of the maxima and relative intensity of the spectral bands, depend on the position and type of substitution. This aspect is key to distinguish positional isomers that contain differences in the substitution around the benzene ring. It is well known that this kind of structure modification is common for cathinones, especially in countries where individual drug control system is used (for example, when 4-MMC was banned, 3-MMC was introduced to the market). The examples are shown in Fig. 4.5. A study of ten sets of positional isomers showed that it was possible to distinguish most of the isomers within a set (Rowe et al. 2017). It is assumed that HPLC-DAD provides information complementary to EI-MS, where the latter technique cannot distinguish between these types of positional isomers. Another advantage of HPLC-DAD is the repeatability and stability of analytical signal. Moreover, as in the analysis of seized materials samples may be dissolved to a large extent, a potential influence of a sample matrix could be neglected. These features make HPLC-DAD a good choice for quantitative analysis. The main disadvantages are instability of retention times and lower separation capacity in comparison to GC, which makes chromatographic peaks in LC broader. This limits the application of HPLC-DAD in the identification process. It should be noted that use of monolithic column and thermostat improves the HPLC-DAD repeatability.

FTIR is an analytical method that identifies substances and their impurities, and specifies the types of bonds in molecules and the substituent effects. IR spectra are very complex and it is very rare for two compounds to have identical spectra in the entire range 600–4000 cm^{-1}, which is a very desirable feature for identification. The commercial libraries are also formed for the FTIR spectra, e.g. Spectral Database for Organic Compounds SDBS, but they contain a limited number of synthetic cathinones. Much more useful are noncommercial, expert libraries, including the FTIR database formed in the RESPONSE project financed by the European Union, which is available online. The shape of FTIR spectrum depends on the vibrational and rotational energy of chemical bonds in a molecule, and

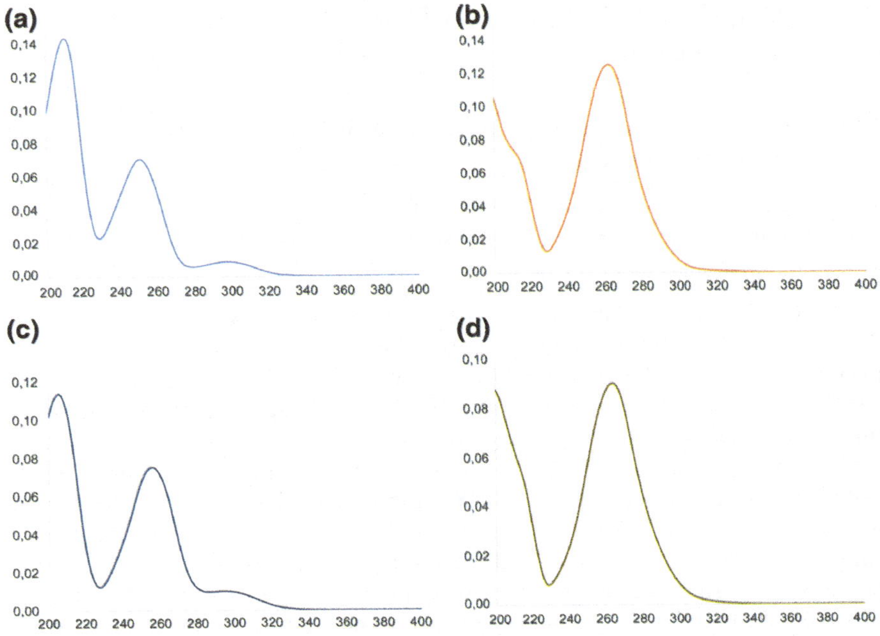

Fig. 4.5 The UV/VIS spectra of: **a** 3-CMC, **b** 4-CMC, **c** 3-MEC and **d** 4-MEC

therefore on the basis of group frequencies (characteristic vibrations) it is possible to determine which functional groups are present in the analysed compounds, which is very useful in the identification process, particularly in distinguishing between positional isomers. The main disadvantage of this method is that the IR spectra of mixtures are complex, and this makes analysis difficult. The measurements may be carried out by the attenuated total reflectance (ATR) technique, so there is no need to pretreat the powders (tablets are only crushed). To analyse small samples, the spectrometer have to be equipped with a microscope.

An ultra-compact, handheld near-infrared (NIR) spectrometer for the classification of 140 different substances, including synthetic cathinones, was presented by (Pederson et al. 2014). They created a library and analysed a set of 25 unknown samples showing the ability to correctly identify the unknown with a very low rate of misidentification, and the ability to use the same library with multiple instruments. It was also proved that through the use of innovative chemometric algorithms, it is possible to identify the individual compounds in an unknown mixture based on the spectral library of the individual compounds only.

High-resolution NMR spectroscopy is a basic technique used for confirmation of organic compounds' structure. Mainly proton (^{1}H) and carbon (^{13}C) spectra are measured, and depending on the chemical composition of a nucleus, these types are supported by ^{15}N, ^{17}O or ^{19}F NMR. In the last years, the role of the 2D ^{1}H–^{13}C spectra in identification has been growing. In the classical approach to structure elucidation by spectral interpretation, reference tables of corresponding analyte values, chemical shifts and spin–spin coupling constants accounting for signal multiplicity are used. The spectral values are very sensitive to changes in molecular configurations and conformations. In the identification process, a computer simulation is used for prediction of NMR spectra for hypothetical structures, followed by comparison with spectra recorded for analytes. Alternatively, such comparisons can be performed if reference databases comprising of experimental NMR spectra are available (Milman 2011). The limitations of NMR application in the analysis of synthetic cathinones are relatively low sensitivity and low identification power in relation to individual components of complex mixtures, therefore most studies of cathinones were performed on pure compounds.

4.3 Analysis of Synthetic Cathinones in Biological Material

4.3.1 Immunochemical Detection

Generally, as screening methods, immunoassays are commonly used for the presumptive identification of drugs of abuse. When a positive result occurs, confirmatory methods, mostly chromatographic are applied. GC-MS and LC-MS are more sensitive and specific, and are used both for qualitative and quantitative analyses. Immunoassay methods are quick as well as cost-efficient and provide information about the drug content in biological specimens, such as blood or urine. Immunoassay tests do not allow the identification of specific compounds but they are based on group identification. The tests require low sample volumes (blood or urine, up to several dozen μL), while providing rapid results. Most commercially available immunoassays are limited to a set number of drugs. Newly emerging compounds are either not or poorly detected by existing immunoassay methods. Despite the fact that structures of many of the synthetic cathinones are closely related to amphetamine, methamphetamine and MDMA, the commercial immunoassays have poor or no cross-reactivity towards NPS. Swortwood et al. (2014) showed that popular immunoassay classic drug tests are mostly not effective for NPS. Other studies indicate some cross-reactivity in some immunoassays, however, the concentrations of cathinones detected (10–100 μg/mL) make the uselessness of these tests in the analysis of most real samples (Regester et al. 2015).

Currently, immunoassay tests for the detection of cathinones are also available. Randox Toxicology introduced the Mephedrone ELISA kit that is sensitive, among others, for mephedrone, methcathinone, 4-fluoromethcathinone (4-FMC; flephedrone), 3-fluoromethcathinone (3-FMC), and a number of additional metabolites. The same company offers MDPV ELISA kit that targets 3,4-methylenodioxypyrovalerone (MDPV), 3,4-methylenedioxy-α-pyrrolidinobutyrophenone (MDPBP) and naphyrone, as well as α-pyrovalerone (α-PVP)/MDPV kit that is sensitive for the same compounds and moreover for pyrovalerone and α-pyrrolidinopentiophenone (Ellefsen et al. 2014). Such ELISA tests are also manufactured by Neogen (Methcathinone kit). In all the abovementioned immunoassays, the cross-reactivity for some compounds is worse, what makes one to suspect that the detection of some cathinones can be difficult and even impossible, leading to false negative results. However, the experiments of Roda and coworkers on Mephedrone/Methcathinone and MDPV ELISA assays show no false negatives and good overall match between data obtained by the ELISA and LC-MS methods, showing only disagreement in few cases concerning positive results (Roda et al. 2016). It is worth to note that studies done by Swortwood and coworkers indicate that the Mephedrone/Methcathinone Randox kit has no cross-reactivity with other amphetamine-like compounds, and therefore may be useful for detection of cathinones in postmortem specimens, without the usual interference from putrefactive amines formed during decomposition (Swortwood et al. 2014). Notwithstanding, the high number of false positive results can also be a problem in immunoassay screening. Ellefsen et al. (2014) conducted a full validation on the Randox Drugs of Abuse V Biochip Array Technology immunoassay for synthetic cathinones in urine. They also screened over 20 thousand authentic specimens and confirmed positives by LC-MS/MS. The confirmation rate of presumptive positive specimens was very low. However, there may be many explanations of such results, including not only poor value of tests but, for example, the presence of metabolites or other cathinones that were not covered by LC-MS/MS method and could be detected by immunoassay. Another explanation, and a very important issue, is the instability of some synthetic cathinones in blood or urine samples (Johnson and Botch-Jones 2013; Sørensen 2011).

Immunoassay tests can be a useful and rapid tool for screening analysis of biological material but, using such tests, it is never certain whether they will detect new substances for which the cross-reactivity has not been tested. Moreover, changes on the drug market (cathinones that become controlled are immediately replaced by new uncontrolled substances) makes that cross-reactivity for many new cathinone derivatives remain unknown, particularly for the newest compounds. Introducing new immunoassay test takes time and nobody can be sure that the detection capabilities of such tests are still valid for newest compounds. This long lead time to optimize new commercial immunoassays for newly introduced synthetic cathinones causes that chromatographic methods are better suited to develop rapid screening method and detect novel designer drugs.

4.3.2 *Chromatographic Methods*

Only modern coupled techniques, especially chromatographic with mass detection, have the features that enable a comprehensive analysis of a wide range of synthetic cathinones in biological material. They provide low limits of detection and unambiguous identification. Many methods for the determination of synthetic cathinones in biological material have been described in the literature. GC-based methods are limited to nonpolar, volatile and thermally stable compounds, otherwise, the derivatization step is required. This is especially important because synthetic cathinones may undergo thermal degradation during GC analysis (Kerrigan et al. 2016). LC-MS methods provide the sensitivity, selectivity, and specificity necessary for screening analysis for many synthetic cathinones, without the need for derivatization. Therefore, LC-MS are preferred and become technique of choice, mainly LC-MS/MS. Many GC-MS and LC-MS conditions for synthetic cathinones analysis in biological materials were summarized by Namera et al. (2015) and Ellefsen et al. (2016).

4.3.2.1 Gas Chromatography–Mass Spectrometry Detection

Several confirmation and screening methods utilizing GC-MS for the determination of synthetic cathinones in biological materials were published (e.g. Meyer et al. 2010a, b; Ojanpera et al. 2011; Strano-Rossi et al. 2010; Saito et al. 2013; Urlates et al. 2014). However, these methods are rarely used in the analysis of biological material as well as cover up to few compounds. This is mainly due to much less sensitive nature of this technique and the associated need to carry out the derivatization process.

Grapp et al. (2017) analysed urine and serum samples for MDPV by GC-MS. Urine (5 mL) was hydrolysed and then extracted by LLE followed by derivatization with pyridine and acetic anhydride. Alternatively, trimethylsilylation was done with MSTFA. Serum samples (1 mL) were prepared by SPE. It is worth pointing out that elaborated method allowed detection of not only parent compound but also its metabolites. Uralets et al. (2014) applied GC-MS method for studying nearly 35 thousand of urine specimens for designer stimulants. After LLE and trifluoroacetylation, samples were screened for drugs and metabolites excreted free in urine. TFAA was also used for derivatization (after LLE of 1 mL of blood) for isolation of MDPBP by Wiergowski et al. (2016). The limits of detection (LOD) of abovementioned methods were in the range of 1.7 ng/mL (serum) to 10 ng/mL (blood). Hong et al. (2016) proposed a GC-MS method to analyse and quantify in urine samples six synthetic cathinones: mephedrone, methylone, butylone, ethylone, pentylone and MDPV. The urine samples were subjected to SPE and derivatization prior to analysis. The MS worked in selective ion monitoring (SIM mode) achieving LOD in the range of 5–20 ng/mL.

4.3.2.2 High-Performance Liquid Chromatography with Diode Array Detection

The HPLC-DAD (or UHPLC-DAD) method, in contrast to GC-MS, does not require the derivatization step. However, due to the worse specificity of the diode array detection, the use of this technique for the analysis of synthetic cathinones in biological material is limited. Simultaneous determination of 4-substituted cathinones (4-MMC, 4-MEC and 4-FMC) in urine by HPLC-DAD was demonstrated by Mayer et al. (2013). Urine samples (0.5 mL) were prepared by SPE. The diode array detection was performed at 262 nm.

4.3.2.3 Liquid Chromatography–Mass Spectrometry Detection

LC-MS and LC-MS/MS methods are used more frequently for analysis of biological material. Many cathinones were identified in biological samples and methods covering single compound were published. Currently, several LC-MS and LC-MS/MS (as well as UHPLC-MS) screening and confirmation methods also exist for the detection of large groups of synthetic cathinones (and metabolites) in biological samples.

Synthetic cathinones were analysed in ante- and postmortem samples. Analysed materials included the whole blood, dried blood spots (DBS), serum, plasma, urine (including dried urine), hair and oral fluids. Some authors analysed also vitreous humour, cerebrospinal fluid and bile (Marinetti and Antonides 2013). The volumes of samples were dependent on the type of the material, and ranged from 0.1 to 2 mL for blood/serum/plasma, and from 0.05 to 2.5 mL for urine. Oral fluid samples (0.4 mL) were collected with the use of different collection devices, e.g. Quantisal (Miller et al. 2017). Hair samples weighed 20–30 mg. With more advanced ways of samples' preparation, it was possible to reduce the volume of samples. Mercolini et al. (2016) employed volumetric absorptive microsampling (VAMS) for preparation of samples. These authors analysed dried samples for the presence of methylone, ethylone, butylone, mephedrone, 4-MEC and MDPV. The advantage of this procedure was an extremely low sample volume (10 μL). DBS in other methods were also produced from 10 μl of the whole blood (Ambach et al. 2014). Another interesting technique used in the analysis of cathinones and their metabolites was DART-MS (LaPointe et al. 2015). In this case, 3 μL samples were spotted onto QuickStrip Sample Cards.

Developed methods included up to several dozen substances. Ammann et al. (2012) reported the use of LC-MS/MS in simultaneous identification of 25 cathinones in the blood. The most robust and universal methods have been capable of detecting and identifying less than 150 compounds. The method developed by Adamowicz and Tokarczyk (2016) allows the simultaneous screening of 143 compounds, including 36 cathinones. Urine analysis is particularly specific due to the fact that target compounds are not only unchanged compounds but also their metabolites. Therefore, methods of analysis of synthetic cathinones in urine covered up to several dozen parent compounds as well as many metabolites (Bertol et al. 2014; Concheiro et al. 2013; Tang et al. 2014).

Most researchers use LLE for isolation from the whole blood, serum and plasma (Ammann et al. 2012). Synthetic cathinones were also isolated from blood using PP (mainly with acetonitrile) and SPE (Adamowicz and Tokarczyk 2016; Ambach et al. 2015; Sørensen 2011, Swortwood et al. 2013; Vaiano et al. 2016). Sometimes other steps, for example ultrafiltration of extracts, are also applied. Less popular are other extraction techniques. Montesano et al. (2017) developed a method to determine simultaneously 16 NPS in the whole blood, including four cathinones, in which blood samples were extracted by miniaturized solid phase extraction (μ-SPE) technique. Odoardi et al. (2015) developed a method of simultaneous extraction of analytes of different classes of NPS (a total of 78 analytes, including 20 cathinones) from blood samples by a dispersive liquid–liquid microextraction (DLLME) technique that involves very small amounts of solvents.

Urine preparation included simple dilution, PP, LLE, SPE, supported liquid extraction (SLE), and salting-out liquid–liquid extraction (SALLE) (Al-Saffar et al. 2013; Ambach et al. 2015; Bell et al. 2011; Concheiro et al. 2013; Glicksberg et al. 2016; Paul et al. 2014; Roda et al. 2016). Urine preparation often required a hydrolysis step, mainly enzymatic with β-glucuronidase (Bertol et al. 2014; Tang et al. 2014). LaPointe et al. (2015) presented a method of analysis of six cathinones and six metabolites in urine directly without any sample preparation and applied solid phase microextraction (SPME).

Synthetic cathinones were isolated from oral fluid by PP (e.g. with acetonitrile), LLE or SPE (Amaratunga et al. 2013; de Castro et al. 2014; Miller et al. 2017). Hair required different preparation procedures. The first step was always decontamination. Procedure presented by Salomone et al. (2016) was very simple and consisted only of washing and incubation. But after decontamination, hair samples can also be grinded, digested or ultrasonificated (e.g. in ethanol). These steps can be followed by LLE or SPE (Alvarez et al. 2017; Lendoiro et al. 2017; Salomone et al. 2016).

The time needed for sample preparation depends on the preparation method, and the chromatographic analysis time was usually from a few to a dozen or so minutes. The fastest screening method for synthetic cathinones in urine was presented by Neifeld et al. (2016). The method utilized the RapidFire tandem mass spectrometry system (RF-MS-MS) that provided a rapid and highly specific screen with minimal sample preparation and a very short analysis time per sample. Presented screening method detected 14 synthetic cathinones in urine with a run time of only 12.6 s. Sample preparation and hydrolysis were performed in a 96-well plate with one MRM transition used for the identification of each compound (Neifeld et al. 2016).

LOD in methods involving blood/serum/plasma were in the ranges of 0.5–3 ng/mL, in urine 0.5–50 ng/mL, in oral fluid 0.025–1 ng/mL, while in hair 0.05–20 pg/mg. Such limits are sufficient for a proper detection, identification and determination of synthetic cathinones in the aforementioned biological materials.

Screening procedures using low-resolution MS are mainly targeted screening procedures. These methods are focused on the quantitative determination of a predefined set of analytes in biological matrices. Almost all methods use MRM and positive ionization modes. MRM provides unique fragment (product) ions, obtained from one (or more) precursor ions that can be monitored and quantified even in a very complicated matrix. Such methods are sensitive and specific, but their main drawback is that target compounds must be known in advance and compounds not

Table 4.1 Examples of the MRM pairs for selected cathinone derivatives along with mass spectrometer parameters (fragmentor voltages and collision energies)

Name	Precursor ion	Product ion	Fragmentor voltage (V)	Collision energy (V)
2,3-DMEC (2,3-Dimethylethcathinone)	206.2	188.2	56	8
		159.2		16
		158.2		32
2,4-DMEC (2,4-Dimethylethcathinone)	206.2	115.1	68	60
		91.2		60
		72.2		12
3,4-DMMC (3,4-Dimethylmethcathinone)	192.1	174.1	85	8
		159.1		20
3-CMC (3-Chloromethcathinone)	198.1	180.1	25	8
		145.1		16
		144.1		32
4-CMC (4-Chloromethcathinone)	198.1	180.0	83	8
		145.1		16
		144.1		36
3-EMC (3-Ethylmethcathinone)	192.1	91.2	58	40
		77.1		52
		65.2		60
3-FMC (3-Fluoromethcathinone)	182.1	164.1	93	8
		149.0		20
		148.0		36
3-MEC (3-Methylethcathinone)	192.1	174.2	62	8
		145.2		16
		144.2		28

(continued)

Table 4.1 (continued)

Name	Precursor ion	Product ion	Fragmentor voltage (V)	Collision energy (V)
4-BMC (Brephedrone, 4-Bromomethcathinone)	242	145.1	85	12
		144.0		36
		77.1		60
4-CEC (4-Chloromethcathinone)	212.1	194.1	86	8
		144.1		28
		77.1		60
4Cl-α-PVP (4-Chloro-α-pyrrolidinovalerophenone)	266.1	125.0	102	24
		111.0		48
		74.1		132
4-EEC (4-Ethylethcathinone)	206.2	188.2	25	8
		159.1		16
		144.1		28
4-EMC (4-Ethylmethcathinone)	192.1	174.1	68	8
		145.1		20
		144.1		36
4-FMC (Flephedrone, 4-Fluoromethcathinone)	182.1	164.1	83	8
		149.0		20
		148.0		36
4-MBC (Benzedrone)	254.2	236.1	87	8
		91.1		24
		65.1		60
4-MDMC (4-Methyldimethcathinone)	192.1	91.2	54	40
		77.2		60
		72.2		28

(continued)

Table 4.1 (continued)

Name	Precursor ion	Product ion	Fragmentor voltage (V)	Collision energy (V)
4-MEAP (NEMNP, 4-Methyl-α-ethylaminopentiophenone)	220.2	202.1	78	8
		144.1		32
		91.1		48
4-MEC (4-Methylethcathinone)	192.1	174.1	143	8
		145.1		16
		91.1		36
4-MeMABP (4-Methylbuphedrone)	192.1	174.2	56	8
		145.2		20
		144.2		32
4-MPD (4-Methylpentedrone)	206.2	188.1	80	8
		144.1		36
		77.1		60
4-MPHP (PV-4, 4′-Methyl-α-pyrrolidinohexiophenone)	260.2	140.1	104	24
		105.1		20
		91.1		48
α-PVP (α-Pyrrolidinopentiophenone)	232.2	126.1	85	24
		91.0		24
		77.0		48
α-PBP (α-Pyrrolidinobutiophenone)	218.2	112.1	83	24
		91.1		20
		77.1		48
α-PPP (α-Pyrrolidinopropiophenone)	204.1	133.0	120	16
		105.1		24
		98.1		28

(continued)

Table 4.1 (continued)

Name	Precursor ion	Product ion	Fragmentor voltage (V)	Collision energy (V)
α-PVT (α-Pyrrolidinopentiothiophenone)	238.1	126.1	41	20
		111.0		36
		97.0		20
bk-DMBDB (Dibutylone)	236.1	161.0	145	16
		86.1		24
		65.1		56
bk-MDDMA (Dimethylone)	222.1	147.0	83	16
		91.1		36
		72.1		16
BMDP (4-Methylenedioxy-*N*-benzylcathinone)	284.1	266.1	87	8
		91.0		24
		65.1		60
Buphedrone	178.1	160.1	85	4
		91.1		16
		77.1		48
Butylone	222.1	204.1	91	4
		174.1		12
		131.0		32
Ephedrone	164.1	146.0	85	8
		131.0		16
		77.0		52
Ethylone	222.1	204.1	89	4
		174.1		12
		146.1		24

(continued)

Table 4.1 (continued)

Name	Precursor ion	Product ion	Fragmentor voltage (V)	Collision energy (V)
Ethcathinone	178.1	160.1	89	8
		131.0		16
		117.0		28
Eutylone	236.1	218.1	89	8
		188.1		16
		174.0		32
MDPBP (3′,4′-Methylenedioxy-α-pyrrolidinobutyrophenone)	262.2	161.1	118	16
		112.1		24
		65.1		60
MDPPP (3′,4′-Methylenedioxy-α-pyrrolidinopropiophenone)	248.1	147.0	91	20
		98.1		24
		91.1		48
MDPV (Methylenedioxypyrovalerone)	276.2	175.1	124	16
		135.0		20
		126.1		24
MeBP (Methylbuphedrone)	191.2	105.1	99	20
		79.1		40
		77.1		52
Mephedrone/2-MMC/3-MMC	178.1	160.1	87	8
		145.1		20
		77.1		56
MePPP (4′-Methyl-α-pyrrolidinopropiophenone)	218.2	119.2	54	24
		98.2		28
		91.2		40

(continued)

Table 4.1 (continued)

Name	Precursor ion	Product ion	Fragmentor voltage (V)	Collision energy (V)
Metamfepramone (Dimethylcathinone)	178.1	133.0	85	12
		105.1		20
		72.1		24
Methedrone (4-Methoxymethcathinone)	194.1	176.1	85	8
		161.0		16
		146.0		28
Mexedrone	208.1	158.2	56	8
		119.1		20
		91.2		40
MDMC (Methylone)	208.1	190.0	91	4
		160.0		12
		132.0		24
MPBP (4′-Methyl-α-pyrrolidinobutiophenone)	232.2	161.1	85	12
		105.1		24
		91.1		48
MPHP (4′-Methyl-α-pyrrolidinohexiophenone)	260.2	140.2	25	28
		105.1		24
		91.1		52
Naphyrone	282.2	211.1	126	12
		155.0		24
		141.0		20
NEB (N-Ethylbuphedrone)	192.1	174.1	68	8
		130.0		32
		91.1		28

(continued)

Table 4.1 (continued)

Name	Precursor ion	Product ion	Fragmentor voltage (V)	Collision energy (V)
NEH (N-Ethylhexedrone)	220.2	130.1	84	36
		91.1		32
		77.1		60
N-Ethylpentylone (Ephylone)	250.1	232.1	84	8
		202.2		16
		174.1		32
Pentedrone (α-Methylaminovalerophenone)	192.1	174.1	89	8
		132.1		16
		91.1		24
Pentylone (Methylenedioxypentedrone)	236.1	218.1	62	8
		188.1		16
		175.1		20
PV-7 (α-PHP, α-Pyrrolidinohexiophenone)	246.2	140.2	54	24
		91.2		24
		77.2		60
PV-8 (α-PHPP)	260.2	154.2	5	28
		91.1		24
		77.1		60

included in the method are undetected, even if present at high concentrations. Table 4.1 shows examples of the MRM pairs along with mass spectrometer parameters (fragmentor voltages and collision energies) for selected cathinone derivatives.

Although most of the developed methods for detection of synthetic cathinones involve monitoring of specific MRM pairs for the analysed compounds, such methods have several disadvantages. In this context, it seems that the future belongs to high-resolution MS (Meyer and Maurer 2016). The use of HRMS has a potential in identification of drugs for which no reference standards are available. This technique is capable of separating mass fragments at the fourth or fifth decimal place. This enables identification of individual compounds based on their exact mass. Moreover, such screening methods allow for acquisition of nontargeted compounds and retrospective data analysis.

The liquid chromatography-high-resolution mass spectrometry (LC-HRMS) and liquid chromatography-high-resolution multiple mass spectrometry (LC-HRMS/MS) were used for analysis of single cathinones and metabolites (Bertol et al. 2014) as well as for the simultaneous determination of a large number of cathinones and metabolites (Concheiro et al. 2013, 2015; Paul et al. 2014). Montesano et al. (2016) developed a LC-HRMS method for a broad screening of NPS in plasma. Data acquisition was in MS/MS and full-scan modes, and allowed the screening of additional substances by a post-run library matching (an in-house database contained over 300 NPS and known metabolites).

4.3.3 Other Techniques

Other methods of determination of cathinones in biological material are less commonly used. Baciu et al. (2016) applied capillary electrophoresis (CE) combined in-line with SPE for an enantioselective determination of cathinone derivatives in human hair. These authors developed and validated method for the chiral separation and determination of *R,S*-mephedrone and one of its metabolites, *R,S*-4-methylephedrine, and *R,S*-methylenedioxypyrovalerone (*R,S*-MDPV) in human hair samples (100 mg). It should be noted that CE can be an alternative to LC methods using chiral columns due to lower costs as enantioselective CE separations are achieved just by dissolving a chiral selector such as a cyclodextrins in the CE buffer. Sample pretreatment procedure in the presented method consisted of washing and pressurized liquid extraction. The used CE instrument was equipped with a DAD and the analytes were detected at 200 nm.

References

Adamowicz P, Tokarczyk B (2016) Simple and rapid screening procedure for 143 new psychoactive substances by liquid chromatography-tandem mass spectrometry. Drug Test Anal 8(7):652–667. https://doi.org/10.1002/dta.1815

Al-Saffar Y, Stephanson NN, Beck O (2013) Multicomponent LC-MS/MS screening method for detection of new psychoactive drugs, legal highs, in urine—experience from the Swedish population. J Chromatogr B 930:112–120. https://doi.org/10.1016/j.jchromb.2013.04.043

Alvarez JC, Etting I, Abe E et al (2017) Identification and quantification of 4-methylethcathinone (4-MEC) and 3,4-methylenedioxypyrovalerone (MDPV) in hair by LC-MS/MS after chronic administration. Forensic Sci Int 270:39–45. https://doi.org/10.1016/j.forsciint.2016.11.028

Amaratunga P, Lorenz Lemberg B, Lemberg D (2013) Quantitative measurement of synthetic cathinones in oral fluid. J Anal Toxicol 37(9):622–628. https://doi.org/10.1093/jat/bkt080

Ambach L, Hernández Redondo A, König S et al (2014) Rapid and simple LC-MS/MS screening of 64 novel psychoactive substances using dried blood samples. Drug Test Anal 6(4):367–375. https://doi.org/10.1002/dta.1505

Ambach L, Redondo AH, König S et al (2015) Detection and quantification of 56 new psychoactive substances in whole blood and urine by LC-MS/MS. Bioanalysis 7(9):1119–1136. https://doi.org/10.4155/bio.15.48

Ammann D, McLaren JM, Gerostamoulos D et al (2012) Detection and quantification of new designer drugs in human blood: part 2—designer cathinones. J Anal Toxicol 36(6):381–389. https://doi.org/10.1093/jat/bks049

Armenta S, Garrigues S, de la Guardia M et al (2015) Detection and characterization of emerging psychoactive substances by ion mobility spectrometry. Drug Test Anal 7(4):280–289. https://doi.org/10.1002/dta.1678

Baciu T, Borrull F, Calull M et al (2016) Enantioselective determination of cathinone derivatives in human hair by capillary electrophoresis combined in-line with solid-phase extraction. Electrophoresis 37(17-18):2352–2362. https://doi.org/10.1002/elps.201600149

Bell C, George C, Kicman AT et al (2011) Development of a rapid LC-MS/MS method for direct urinalysis of designer drugs. Drug Test Anal 3(7–8):496–504. https://doi.org/10.1002/dta.306

Bertol E, Mari F, Boscolo Berto R et al (2014) A mixed MDPV and benzodiazepine intoxication in a chronic drug abuser: determination of MDPV metabolites by LC-HRMS and discussion of the case. Forensic Sci Int 243:149–155. https://doi.org/10.1016/j.forsciint.2014.08.002

Concheiro M, Anizan S, Ellefsen K et al (2013) Simultaneous quantification of 28 synthetic cathinones and metabolites in urine by liquid chromatography-high resolution mass spectrometry. Anal Bioanal Chem 405:9437–9448. https://doi.org/10.1007/s00216-013-7386-z

Concheiro M, Castaneto M, Kronstrand R et al (2015) Simultaneous determination of 40 novel psychoactive stimulants in urine by liquid chromatography-high resolution mass spectrometry and library matching. J Chromatogr A 1397:32–42. https://doi.org/10.1016/j.chroma.2015.04.002

de Castro A, Lendoiro E, Fernández-Vega H et al (2014) Liquid chromatography tandem mass spectrometry determination of selected synthetic cathinones and two piperazines in oral fluid. Cross reactivity study with an on-site immunoassay device. J Chromatogr A 1374:93–101. https://doi.org/10.1016/j.chroma.2014.11.024

Ellefsen KN, Anizan S, Castaneto MS et al (2014) Validation of the only commercially available immunoassay for synthetic cathinones in urine: randox drugs of abuse V biochip array technology. Drug Test Anal 6(7-8):728–738. https://doi.org/10.1002/dta.1633

Ellefsen KN, Concheiro M, Huestis MA (2016) Synthetic cathinone pharmacokinetics, analytical methods, and toxicological findings from human performance and postmortem cases. Drug Metab Rev 48(2):237–265. https://doi.org/10.1080/03602532.2016.1188937

Glicksberg L, Bryand K, Kerrigan S (2016) Identification and quantification of synthetic cathinones in blood and urine using liquid chromatography-quadrupole/time of flight (LC-Q/

TOF) mass spectrometry. J Chromatogr B Analyt Technol Biomed Life Sci 1035:91–103. https://doi.org/10.1016/j.jchromb.2016.09.027

Grapp M, Kaufmann C, Ebbecke M (2017) Toxicological investigation of forensic cases related to the designer drug 3,4-methylenedioxypyrovalerone (MDPV): detection, quantification and studies on human metabolism by GC-MS. Forensic Sci Int 273:1–9. https://doi.org/10.1016/j.forsciint.2017.01.021

Gwak S, Arroyo-Mora LE, Almirall JR (2015) Qualitative analysis of seized synthetic cannabinoids and synthetic cathinones by gas chromatography triple quadrupole tandem mass spectrometry. Drug Test Anal 7(2):121–130. https://doi.org/10.1002/dta.1667

Hong WY, Ko YC, Lin MC et al (2016) Determination of synthetic cathinones in urine using gas chromatography-mass spectrometry techniques. J Anal Toxicol 40(1):12–16. https://doi.org/10.1093/jat/bkv108

Johnson RD, Botch-Jones SR (2013) The stability of four designer drugs: MDPV, mephedrone, BZP and TFMPP in three biological matrices under various storage conditions. J Anal Toxicol 37(2):51–55. https://doi.org/10.1093/jat/bks138

Joshi M, Cetroni B, Camacho A et al (2014) Analysis of synthetic cathinones and associated psychoactive substances by ion mobility spectrometry. Forensic Sci Int 244:196–206. https://doi.org/10.1016/j.forsciint.2014.08.033

Kerrigan S, Savage M, Cavazos C et al (2016) Thermal degradation of synthetic cathinones: implications for forensic toxicology. J Anal Toxicol 40(1):1–11. https://doi.org/10.1093/jat/bkv099

LaPointe J, Musselman B, O'Neill T et al (2015) Detection of "bath salt" synthetic cathinones and metabolites in urine via DART-MS and solid phase microextraction. J Am Soc Mass Spectrom 26(1):159–165. https://doi.org/10.1007/s13361-014-1006-9

Lendoiro E, Jiménez-Morigosa C, Cruz A et al (2017) An LC-MS/MS methodological approach to the analysis of hair for amphetamine-type-stimulant (ATS) drugs, including selected synthetic cathinones and piperazines. Drug Test Anal 9(1):96–105. https://doi.org/10.1002/dta.1948

Liu C, Jia W, Li T et al (2017) Identification and analytical characterization of nine synthetic cathinone derivatives N-ethylhexedrone, 4-Cl-pentedrone, 4-Cl-α-EAPP, propylone, N-ethylnorpentylone, 6-MeO-bk-MDMA, & α-PiHP, 4-Cl-α-PHP, and 4-F-α-PHP. Drug Test Anal 9(8):1162–1171. https://doi.org/10.1002/dta.2136

Marinetti LJ, Antonides HM (2013) Analysis of synthetic cathinones commonly found in bath salts in human performance and postmortem toxicology: Method development, drug distribution and interpretation of results. J Anal Toxicol 37(3):135–146. https://doi.org/10.1093/jat/bks136

Mayer M, Benko A, Huszár A et al (2013) Simultaneous determination of 4-substituted cathinones (4-MMC, 4-MEC and 4-FMC) in human urine by HPLC-DAD. J Chromatogr Sci 51(9):861–866. https://doi.org/10.1093/chromsci/bms183

Mercolini L, Protti M, Catapano MC et al (2016) LC-MS/MS and volumetric absorptive microsampling for quantitative bioanalysis of cathinone analogues in dried urine, plasma and oral fluid samples. J Pharm Biomed Anal 123:186–194. https://doi.org/10.1016/j.jpba.2016.02.015

Meyer MR, Maurer HH (2016) Review: LC coupled to low- and high-resolution mass spectrometry for new psychoactive substance screening in biological matrices—where do we stand today? Anal Chim Acta 927:13–20. https://doi.org/10.1016/j.aca.2016.04.046

Meyer MR, Du P, Schuster F et al (2010a) Studies on the metabolism of the α-pyrrolidinophenone designer drug methylenedioxypyrovalerone (MDPV) in rat and human urine and human liver microsomes using GC-MS and LC-high-resolution MS and its detectability in urine by GC-MS. J Mass Spectrom 45:1426–1442. https://doi.org/10.1002/jms.1859

Meyer MR, Wilhelm J, Peters FT et al (2010b) Beta-keto amphetamines: studies on the metabolism of the designer drug mephedrone and toxicological detection of mephedrone, butylone, and methylone in urine using gas chromatography-mass spectrometry. Anal Bioanal Chem 397(3):1225–1233. https://doi.org/10.1007/s00216-010-3636-5

Miller B, Kim J, Concheiro M (2017) Stability of synthetic cathinones in oral fluid samples. Forensic Sci Int 274:13–21. https://doi.org/10.1016/j.forsciint.2016.11.034

Milman BL (2011) Chemical identification and its quality assurance. Springer, Berlin Heidelberg. https://doi.org/10.1007/978-3-642-15361-7_2

Montesano C, Vannutelli G, Gregori A et al (2016) Broad screening and identification of novel psychoactive substances in plasma by high-performance liquid chromatography-high-resolution mass spectrometry and post-run library matching. J Anal Toxicol 40(7):519–528. https://doi.org/10.1093/jat/bkw043

Montesano C, Vannutelli G, Piccirilli V et al (2017) Application of a rapid μ-SPE clean-up for multiclass quantitative analysis of sixteen new psychoactive substances in whole blood by LC-MS/MS. Talanta 167:260–267. https://doi.org/10.1016/j.talanta.2017.02.019

Musah RA, Cody RB, Domin MA et al (2014) DART-MS in-source collision induced dissociation and high mass accuracy for new psychoactive substance determinations. Forensic Sci Int 244:42–49. https://doi.org/10.1016/j.forsciint.2014.07.028

Namera A, Kawamura M, Nakamoto A et al (2015) Comprehensive review of the detection methods for synthetic cannabinoids and cathinones. Forensic Toxicol 33(2):175–194. https://doi.org/10.1007/s11419-015-0270-0

Neifeld JR, Regester LE, Holler JM et al (2016) Ultrafast screening of synthetic cannabinoids and synthetic cathinones in urine by RapidFire-Tandem Mass Spectrometry. J Anal Toxicol 40 (5):379–837. https://doi.org/10.1093/jat/bkw025

Nie H, Li X, Hua Z et al (2016) Rapid screening and determination of 11 new psychoactive substances by direct analysis in real time mass spectrometry and liquid chromatography/quadrupole time-of-flight mass spectrometry. Rapid Commun Mass Spectrom 30(S1):141–146. https://doi.org/10.1002/rcm.7629

Odoardi S, Fisichella M, Romolo FS et al (2015) High-throughput screening for new psychoactive substances (NPS) in whole blood by DLLME extraction and UHPLC-MS/MS analysis. J Chromatogr B Analyt Technol Biomed Life Sci 1000:57–68. https://doi.org/10.1016/j.jchromb.2015.07.007

Ojanpera IA, Heikman PK, Rasanen IJ (2011) Urine analysis of 3,4-methylenedioxypyrovalerone in opioid-dependent patients by gas chromatography–mass spectrometry. Ther Drug Monit 33 (2):257–263. https://doi.org/10.1097/FTD.0b013e318208b693

Paul M, Ippisch J, Herrmann C et al (2014) Analysis of new designer drugs and common drugs of abuse in urine by a combined targeted and untargeted LC-HR-QTOFMS approach. Anal Bioanal Chem 406(18):4425–4441. https://doi.org/10.1007/s00216-014-7825-5

Pederson CG, Friedrich DM, Hsiung C et al (2014) Pocket-size near-infrared spectrometer for narcotic materials identification. In: Proceedings of SPIE—the international society for optical engineering, 9101, Art. no. 91010O. https://doi.org/10.1117/12.2050019

Philp M, Shimmon R, Tahtouh M et al (2016) Development and validation of a presumptive colour spot test method for the detection of synthetic cathinones in seized illicit materials. Forensic Chem 1:39–50. https://doi.org/10.1016/j.forc.2016.06.001

Qian Z, Jia W, Li T et al (2017) Identification of five pyrrolidinyl substituted cathinones and the collision-induced dissociation of electrospray-generated pyrrolidinyl substituted cathinones. Drug Test Anal 9(5):778–787. https://doi.org/10.1002/dta.2035

Randox Toxicology—ELISA https://www.randoxtoxicology.com/products/elisa. Accessed 4 Oct 2017

Regester LE, Chmiel JD, Holler JM et al (2015) Determination of designer drug cross-reactivity on five commercial immunoassay screening kits. J Anal Toxicol 39(2):144–151. https://doi.org/10.1093/jat/bku133

Roda E, Lonati D, Buscaglia E et al (2016) Evaluation of two different screening ELISA assays for synthetic cathinones (mephedrone/methcathinone and MDPV) with LC-MS method in intoxicated patients. J Clin Toxicol 6:3. https://doi.org/10.4172/2161-0495.1000302

Rosner P (2017) Mass spectra of designer drugs 2017. Wiley VCH, Weinheim

Rowe WF, Marginean I, Carnes S et al (2017) The role of diode array ultraviolet detection for the identification of synthetic cathinones. Drug Test Anal. https://doi.org/10.1002/dta.2163

Saito T, Namera A, Osawa M et al (2013) SPME–GC–MS analysis of a-pyrrolidinovaleorophenone in blood in a fatal poisoning case. Forensic Toxicol 31(2):328–332. https://doi.org/10.1007/s11419-013-0183-8

Salomone A, Gazzilli G, Di Corcia D et al (2016) Determination of cathinones and other stimulant, psychedelic, and dissociative designer drugs in real hair samples. Anal Bioanal Chem 408 (8):2035–2042. https://doi.org/10.1007/s00216-015-9247-4

Sørensen LK (2011) Determination of cathinones and related ephedrines in forensic whole-blood samples by liquid-chromatography-electrospray tandem mass spectrometry. J Chromatogr B Analyt Technol Biomed Life Sci 879(11–12):727–736. https://doi.org/10.1016/j.jchromb.2011.02.010

Stojanovska N, Tahtouh M, Kelly T et al (2014) Presumptive analysis of 4-methylmethcathinone (mephedrone) using Desorption electrospray ionisation—mass spectrometry (DESI-MS). Aust J Forensic Sci 46(4):411–423. https://doi.org/10.1080/00450618.2013.879206

Strano-Rossi S, Cadwallader AB, de la Torre X et al (2010) Toxicological determination and in vitro metabolism of the designer drug methylenedioxypyrovalerone (MDPV) by gas chromatography/mass spectrometry and liquid chromatography/quadrupole time-of-flight mass spectrometry. Rapid Commun Mass Spectrom 24:2706–2714. https://doi.org/10.1002/rcm.4692

Swortwood MJ, Boland DM, DeCaprio AP (2013) Determination of 32 cathinone derivatives and other designer drugs in serum by comprehensive LC-QQQ-MS/MS analysis. Anal Bioanal Chem 405:1383–1397. https://doi.org/10.1007/s00216-012-6548-8

Swortwood MJ, Hearn WL, DeCaprio AP (2014) Cross-reactivity of designer drugs, including cathinone derivatives, in commercial enzyme-linked immunosorbent assays. Drug Test Anal 6 (7–8):716–727. https://doi.org/10.1002/dta.1489

Synthetic Cathinones (Methcathinone) ELISA Kit. http://www.neogeneurope.com/Toxicology/pdf/ProdInfo/SyntheticCathinones.pdf. Accessed 4 Oct 2017

Tang MHY, Ching CK, Lee CYW et al (2014) Simultaneous detection of 93 conventional and emerging drugs of abuse and their metabolites in urine by UHPLC-MS/MS. J Chromatogr B 969:272–284. https://doi.org/10.1016/j.jchromb.2014.08.033

Toole KE, Fu S, Shimmon RG et al (2012) Color tests for the preliminary identification of methcathinone and analogues of methcathinone. Microgram J 9(1):27–32

Uralets V, Rana S, Morgan S et al (2014) Testing for designer stimulants: metabolic profiles of 16 synthetic cathinones excreted free in human urine. J Anal Toxicol 38(5):233–241. https://doi.org/10.1093/jat/bku021

Vaiano F, Busardò FP, Palumbo D et al (2016) A novel screening method for 64 new psychoactive substances and 5 amphetamines in blood by LC-MS/MS and application to real cases. J Pharm Biomed Anal 129:441–449. https://doi.org/10.1016/j.jpba.2016.07.009

Westphal F, Junge T, Rosner P et al (2009) Mass and NMR spectroscopic characterization of 3,4-methylenedioxypyrovalerone: a designer drug with α-pyrrolidinophenone structure. Forensic Sci Int 190(1-3):1–8. https://doi.org/10.1016/j.forsciint.2009.05.001

Wiergowski M, Woźniak MK, Kata M et al (2016) Determination of MDPBP in postmortem blood samples by gas chromatography coupled with mass spectrometry. Monatsh Chem 47:1415–1421. https://doi.org/10.1007/s00706-016-1780-0

Zuba D (2012) Identification of cathinones and other active components of 'legal highs' by mass spectrometric methods. TrAC Trend Anal Chem 32:15–30. https://doi.org/10.1016/j.trac.2011.09.009

Chapter 5
Metabolism of Synthetic Cathinones

Kei Zaitsu

Abstract Synthetic cathinones have recently emerged and are abused worldwide. To date, numerous cathinone derivatives have been identified as new psychoactive substances, and their chemical structures have gradually become complicated. Accordingly, metabolic pathways of synthetic cathinones differ depending on their chemical structures, and forensic toxicologists are critically required to fully understand their metabolism features in order to prove the intake of such cathinone derivatives. In this chapter, major phase I metabolic pathways of cathinones are classified according to their chemical structures, and the differences in their metabolic profiles are discussed on the basis of in vivo and in vitro metabolism studies. Although there is limited information on phase II metabolism of synthetic cathinones, some characterized pathways such as glucuronidation and sulfation, which are identified not only for their hydrolyzed- and/or hydroxylated-metabolites but also for unchanged cathinone derivatives, are also described.

Keywords Synthetic cathinones · Metabolism · Phase I metabolism
Phase II metabolism

5.1 Chemical Structures of Synthetic Cathinones—A Basis to Understand Their Metabolic Pathways

Prior to discussing about the major metabolic pathways of synthetic cathinones, we must become familiar with their classification; this is because the major metabolic pathways of synthetic cathinones are determined by their chemical structures.

K. Zaitsu (✉)
Department of Legal Medicine and Bioethics, Nagoya University Graduate School of Medicine, 65 Tsurumai-cho, Showa-ku, Nagoya 466-8550, Japan
e-mail: kzaitsu@med.nagoya-u.ac.jp

J. B. Zawilska (ed.), *Synthetic Cathinones*, Current Topics in Neurotoxicity 12,
https://doi.org/10.1007/978-3-319-78707-7_5

Fig. 5.1 Basic structure of synthetic cathinone

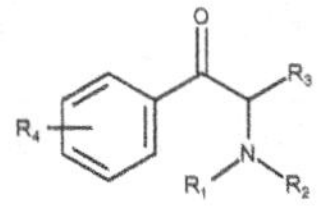

In this section, the synthetic cathinones are classified according to their main chemical structures, though more detailed information on their classification are described in Chap. 6.

As shown in Fig. 5.1, synthetic cathinones have a β-keto-phenethylamine core structure (Zawilska and Wojcieszak 2013), which is closely related to amphetamine; synthetic cathinones are derivatives of amphetamine-type stimulants. Therefore, synthetic cathinones and amphetamine derivatives have some overlapping metabolic pathways. Over the last decade, more than 30 cathinone derivatives have been distributed in illicit drug markets (Valente et al. 2014). According to Valente and coworkers, synthetic cathinones can be classified into four families as follows (Valente et al. 2014):

(1) ***N*-Alkylated cathinone derivatives with/without ring substituents**. In this group, ethcathinone, ephedrone, mephedrone, 3′,4′-dimethylmethcathinone (3,4-DMMC), and so on are classified.
(2) ***N*-Pyrrolidine cathinone derivatives with ring substituents**. Compounds from this group commonly have an *N*-pyrrolidine moiety instead of *N*-alkyl chain. This group includes pyrovalerone, α-PVP, MPHP, and so on.
(3) **3′,4′-Methylenedioxy-*N*-alkylated cathinones**. In this group, 3′,4′-methylenedioxy moiety is added to the benzene ring, like 3,4-methylenedioxy methamphetamine (MDMA). Methylone (bk-MDMA), ethylone (bk-MDEA), butylone (bk-MBDB), and pentylone (bk-MBDP) and some compounds are classified to this group.
(4) **3′,4′-Methylenedioxy-*N*-pyrrolidine cathinone derivatives**. Compounds in this group commonly contain both the 3′,4′-methylenedioxy moiety added to benzene ring as well as the *N*-pyrrolidine moiety. MDPV, MDPPP, and MDPBP are classified to this group.

Main metabolic pathways for these four groups are different because of their chemical properties, though common metabolic pathways are observed intragroup. In the next section, common metabolic pathways are outlined.

5.2 Outline of Major Metabolic Pathways for Synthetic Cathinones

5.2.1 *Phase I Metabolism*

In this section, major phase I metabolic pathways for synthetic cathinones are outlined as shown in Figs. 5.2, 5.3, 5.4, 5.5, 5.6, and 5.7. Almost all synthetic cathinones are excreted as their unchanged form in urine (Kelly 2011; Valente et al.

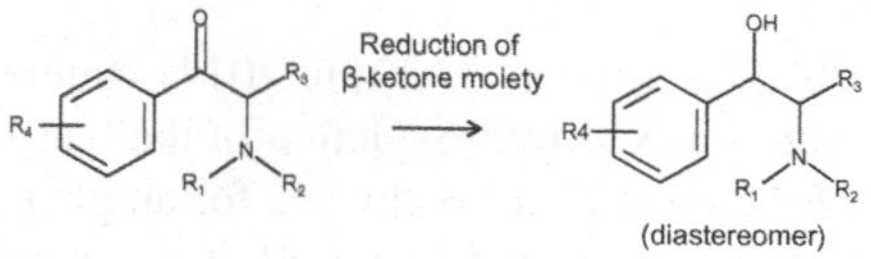

Fig. 5.2 Reduction of β-ketone moiety

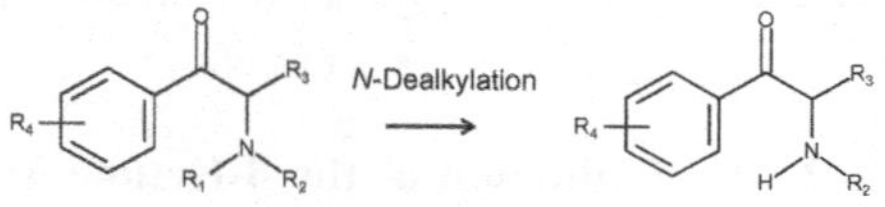

Fig. 5.3 *N*-dealkylation

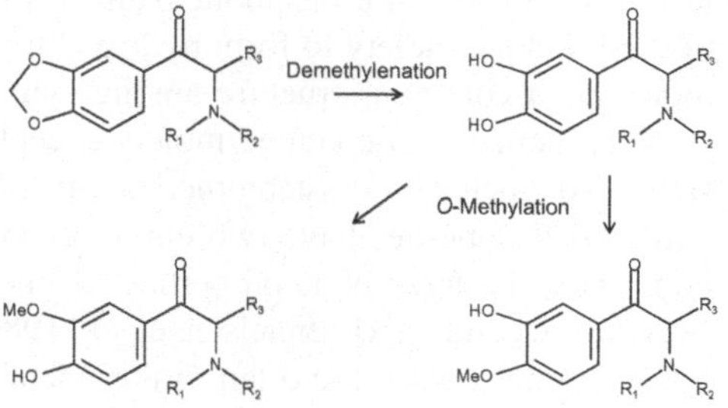

Fig. 5.4 Demethylenation followed by *O*-methylation metabolic pathways

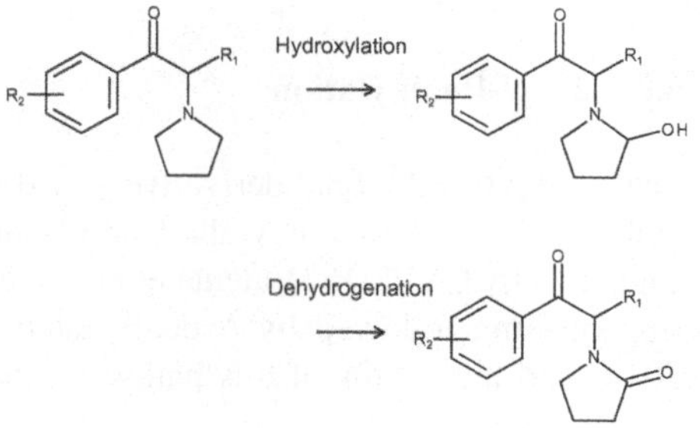

Fig. 5.5 Hydroxylation followed by dehydrogenation, and ring opening for the *N*-pyrrolidine cathinone derivative

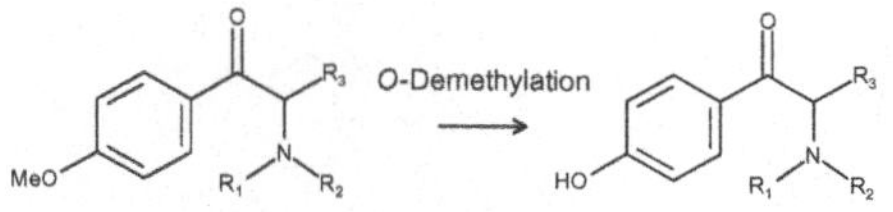

Fig. 5.6 *O*-Demethylation of 4′-methoxy group

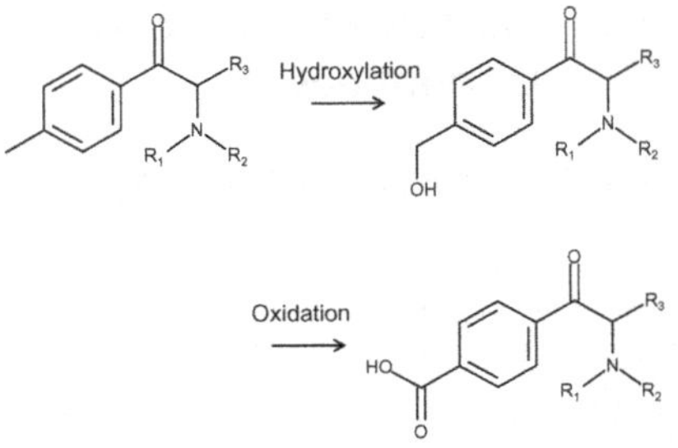

Fig. 5.7 Hydroxylation of the 4′-methyl group followed by oxidation to the corresponding carboxylic acid

2014; Zaitsu et al. 2011, 2014). Interestingly, for non-ring substituted synthetic cathinones, 4-hydroxylation of the benzene ring is almost negligible, while it is one of the most major pathways for amphetamines (Caldwell et al. 1972). Although the reason as to why 4-hydroxylation is rarely observed for synthetic cathinones has not been clarified yet, substitution of the ketone moiety to β-carbon changes not only chemical properties, but also metabolic profiles.

5.2.1.1 Reduction of the β-Ketone Moiety

One of the most major metabolic pathways for synthetic cathinones is the reduction of the β-ketone moiety to form hydroxyl metabolites (Fig. 5.2). Since the β-ketone moiety is a common structure among synthetic cathinones, almost all cathinones show reduction of the ketone moiety except in some special cases. This metabolic step also includes diastereomerization of the generated metabolites because hydroxyl metabolites have two chiral carbons intra-molecule and these chiral carbons show diastereomatic properties. Generation rate of each diastereomer may be structure-dependent (Brenneisen et al. 1986; Zaitsu et al. 2009). The generated hydroxyl metabolites are often further metabolized by phase II metabolism such as glucuronidation and/or sulfation, which will be mentioned later.

5.2.1.2 *N*-Dealkylation

For *N*-alkyl cathinone derivatives, *N*-dealkylation is also the major metabolic pathway as well as for *N*-alkyl amphetamine derivatives (Fig. 5.3); it is mainly catalyzed by CYP2D6 (Valente et al. 2014). For pyrrolidine derivatives, pyrrolidine ring opening followed by *N*-dealkylated pathways is also observed (Zaitsu et al. 2014) and the details of this pathway are described later.

5.2.1.3 Demethylenation Followed by *O*-Methylation Pathways for 3′,4′-Methylenedioxyphenyl Cathinone Derivatives

For 3′,4′-methylenedioxyphenyl cathinone derivatives, demethylenation of the 3′, 4′-methylenedioxy moiety to dihydroxy metabolite, and its *O*-methylation pathway are the major metabolic pathways (Kamata et al. 2006; Zaitsu et al. 2011) (Fig. 5.4). These metabolic pathways are well known for 3′,4′-methylenedioxyamphetamines (MDAs) (Shima et al. 2008). Demethylenation of the 3′,4′-methylenedioxy moiety is generally mediated by CYP2D6 and CYP2C19 (Meyer et al. 2008), while further *O*-methylation of dihydroxy metabolite is mediated by catechol *O*-methyl transferase (COMT) (Zaitsu et al. 2011).

O-methylation is generally more abundant for the 4′-hydroxyl moiety than for the 3′-hydroxyl moiety (Kamata et al. 2006; Zaitsu et al. 2009). The reason why 4′-methoxy-3′-hydroxyl metabolite is abundant is veiled, though it may be due to steric hindrance at the 3′-hydroxyl moiety during the methylation step by COMT.

The generated 4′-methoxy-3′-hydroxyl and 3′-methoxy-4′-hydroxyl metabolites are further metabolized by phase II metabolism such as glucuronidation and/or sulfation as explained later.

5.2.1.4 Hydroxylation Followed by Dehydrogenation, and Ring Opening for the *N*-Pyrrolidine Cathinone Derivatives

For the *N*-pyrrolidine cathinone derivatives, hydroxylation at 2″-position of the *N*-pyrrolidine moiety commonly occurs. The generated 2″-hydroxyl metabolite is further metabolized by dehydrogenation to 2″-oxo-metabolite (lactam) (Valente et al. 2014; Zaitsu et al. 2014). Finally, the pyrrolidine ring of the 2″-oxo-metabolite opens to generate the corresponding aliphatic aldehyde metabolite. The aldehyde metabolite is finally oxidized to carboxylic acid metabolite (Fig. 5.5).

5.2.1.5 Metabolism of Benzene Ring Substituents

As with other ring-substituent derivatives of amphetamines, metabolic reaction of benzene ring substituents is observed. For 4′-methoxyphenyl cathinone derivatives, *O*-demethylation generates the corresponding 4′-hydroxyl-phenyl metabolites (Negreira et al. 2015) (Fig. 5.6). This *O*-demethylation pathway is commonly observed in 4′-methoxyamphetamines (*p*-methoxyamphetamines, PMAs) (Staack and Maurer 2005; Staack et al. 2004; Zaitsu et al. 2008).

4′-Hydroxy metabolites are generally metabolized to the corresponding glucuronide and/or sulfate, as mentioned below. For 4′-methylphenyl cathinone derivatives, hydroxylation of the 4′-methyl moiety of the phenyl ring occurs, and the generated hydroxyl-methyl metabolite is further oxidized to the corresponding carboxylic acid (Fig. 5.7).

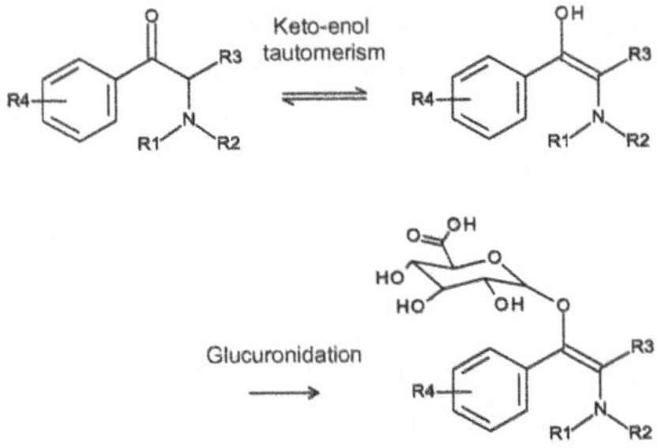

Fig. 5.8 Glucuronidation of the unchanged synthetic cathinones due to keto-enol tautomerism

5.2.2 Phase II Metabolism

Like phase II metabolism for amphetamine derivatives, hydroxyl metabolites of synthetic cathinones generated by phase I metabolism can undergo phase II metabolism. Hydroxyl metabolites generated by reduction of the β-ketone moiety are also metabolized to their corresponding glucuronide and/or sulfate (Matsuta et al. 2015; Shima et al. 2014, 2015; Zaitsu et al. 2009).

Interestingly, there is some evidence that unchanged synthetic cathinone itself can undergo phase II metabolism, especially glucuronidation (Kamata et al. 2006; Shima et al. 2013; Zaitsu et al. 2009). This transformation is due to keto-enol tautomerism: the enol form of synthetic cathinones can be transformed to their conjugates as shown in Fig. 5.8. Therefore, enzymatic hydrolysis using glucuronidase/sulfatase of biological specimens such as urine can sometimes increase the concentration of the unchanged synthetic cathinone (Kamata et al. 2006; Shima et al. 2013; Zaitsu et al. 2009).

5.3 Details of Metabolism for Each Compound

5.3.1 N-Alkylated Cathinone Derivatives with/Without Ring Substituents

5.3.1.1 Cathinone, Methcathinone (*N*-Methylcathinone, Ephedrone), Ethcathinone (*N*-Ethylcathinone), Dimethylpropion (*N*,*N*-Dimethylcathinone) and Diethylpropion (*N*,*N*-Diethylcathinone, Amfepramone)

Cathinone is one of the psychoactive phenylethylamine alkaloids present in the khat plant (*Catha edulis*) (Kelly 2011), and methcathinone, ethcathinone, dimethylpropion and diethylpropion are its *N*-methyl, *N*-ethyl, *N*,*N*-dimethyl and *N*,*N*-diethyl

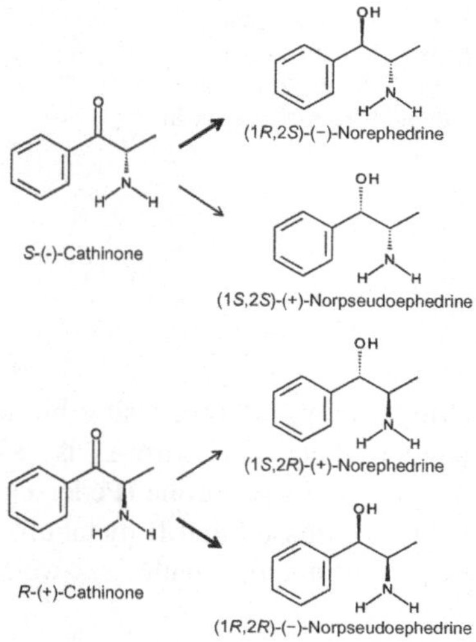

Fig. 5.9 Stereoselective reduction of the β-ketone group of cathinone (adapted from Brenneisen et al. 1986). Bold arrow indicates the more abundant pathway

derivatives. These compounds are commonly metabolized via reduction of the β-ketone group. *N*-dealkylation is also a major metabolic pathway except for cathinone.

Cathinone is mainly metabolized to norephedrine and norpseudoephedrine by reduction of the β-ketone moiety (Brenneisen et al. 1986; Guantai and Maitai 1983; see Chapter 3); stereoselective metabolic profiles are observed (Fig. 5.9). The major metabolite of *S*-(−)-cathinone is norephedrine, whereas the major metabolite of *R*-(+)-cathinone is norpseudoephedrine. Brenneisen and coworkers reported that oral administration of cathinone (racemate form) led to 22–52% of the β-ketone-reduced hydroxyl metabolites after 24 h in human urine (Brenneisen et al. 1986).

Stereoselective metabolism of methcathinone and dimethypropion to the corresponding β-ketone-reduced hydroxyl metabolites were also reported (Paul and Cole 2001; Schreiber et al. 1968; Testa and Beckett 1972, 1973). Methcathinone, ethcathinone, demethylpropion, and diethylpropion are also metabolized by *N*-dealkyl reaction.

Schreiber et al. (1968) demonstrated that the most abundant metabolite of diethylpropion is hippuric acid, which showed 27% of the radioactivity excreted in the urine between 8 and 12 h after ingestion on the basis of the human study using ^{14}C-labeled diethylpropion.

5.3.1.2 4′-Methylmethcathinone (Mephedrone, 4-MMC)

4′-Methylmethcathinone (mephedrone) is excreted as an unchanged form in human and rat urine. Mephedrone is metabolized via *N*-demethylation and hydroxylation of the 4′-methyl moiety of the phenyl ring (Meyer et al. 2010b) (Fig. 5.10). 4′-Hydroxyl-methyl-metabolite is further metabolized to the carboxylic metabolite

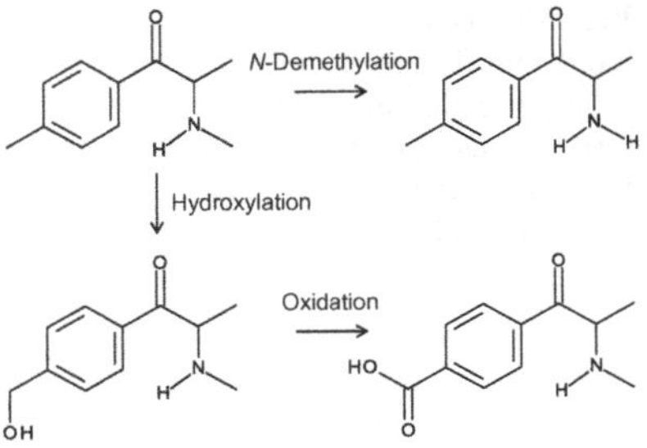

Fig. 5.10 Major phase I metabolic pathways for 4′-methylmethcathinone (adapted from Meyer et al. 2010b)

(Meyer et al. 2010b). Using human liver microsomal preparations, it was demonstrated that mephedrone is also metabolized to dihydro-mephedrone and 4′-carboxy-mephedrone (Pedersen et al. 2013). Khreit et al. (2013) reported on the details of phase I and II metabolism in rat liver hepatocytes, where acetylation and/or glucuronidation pathways were observed as phase II metabolism.

5.3.1.3 4′-Methoxymethcathinone (Methedrone)

Studies on human liver microsomes and human liver cytosol revealed that 4′-methoxymethcathinone (methedrone) is metabolized via *N*-demethylation, *O*-demethylation, and hydroxylation of the aliphatic side chain and reduction of the β-ketone group (Negreira et al. 2015) (Fig. 5.11). It was also demonstrated that 4′-methoxymethcathinone is excreted in postmortem human urine as the unchanged form (Wikström et al. 2010).

5.3.1.4 3′,4′-Dimethylmethcathinone (3,4-DMMC)

Much of the ingested 3′,4′-dimethylmethcathinone (3,4-DMMC) is excreted as an unchanged form; it is also partially conjugated by glucuronidation (Shima et al. 2013).

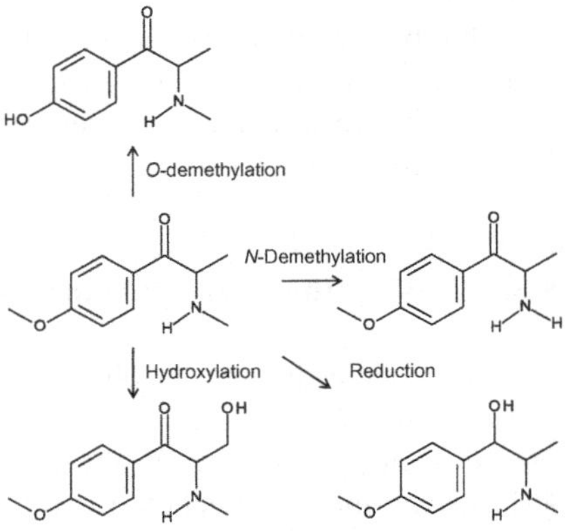

Fig. 5.11 Major phase I metabolic pathways for 4′-methoxymethcathinone (adapted from Negreira et al. 2015)

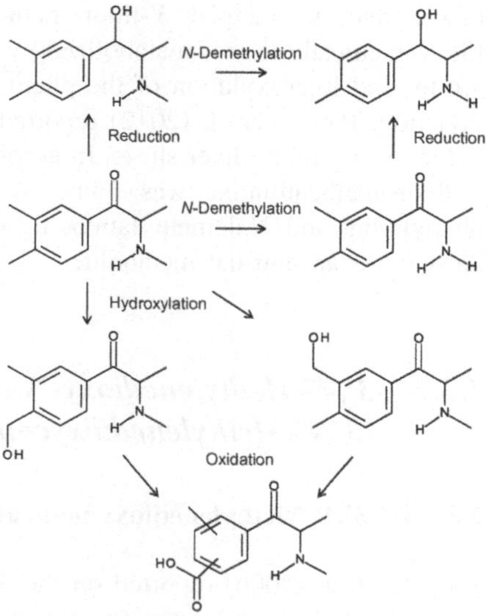

Fig. 5.12 Major phase I metabolic pathways for 3′,4′-dimethylmethcathinone (adapted from Shima et al. 2013)

3,4-DMMC is also metabolized via *N*-demethylation, reduction of the β-ketone group and hydroxylation of the 3′- or 4′-methyl group of the phenyl ring. 3′-Hydroxymethyl and 4′-hydroxymethyl metabolites are further oxidized to the corresponding carboxylic metabolites (Fig. 5.12). In addition, β-ketone-reduced hydroxyl metabolites are partially glucuronidated.

5.3.1.5 3′-Fluoromethcathinone and 3′-Bromomethcathinone

Meyer et al. (2012) described the metabolism of 3′-fluoromethcathinone and 3′-bromomethcathinone in vivo (rat) and in vitro (human liver microsomes). In their study, the unchanged 3′-bromomethcathinone was not detected in rat urine. The compound is metabolized by *N*-demethylation, reduction of the β-ketone group, and hydroxylation of the phenyl ring (Fig. 5.13). The β-ketone-reduced hydroxyl metabolite is further glucuronidated. In rat urine, no sulfate conjugates were detected, though it is unclear whether sulfates exist in human urine or not. On the

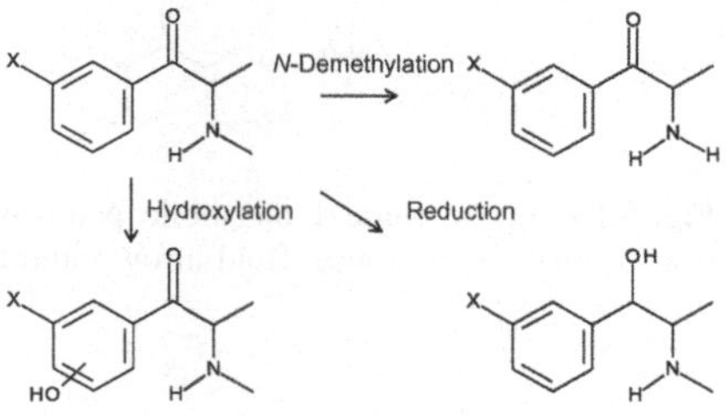

Fig. 5.13 Major phase I metabolic pathways for 3′-fluoromethcathinone and 3′-bromomethcathinone. (*X* = F or Br) (adapted from Meyer et al. 2012)

other hand, unchanged 3′-fluoromethcathinone was excreted in rat urine. 3′-Fluoromethcathinone is metabolized by *N*-demethylation, reduction of the β-ketone group, and hydroxylation of the phenyl ring. CYP2B6 mainly mediates *N*-demethylation. Pawlik et al. (2012) reported the 3′-fluoromethcathinone phase I metabolism using rabbit liver slices. In accordance with the data of Meyer et al. (2012), 3′-fluoromethcathinone was found to be metabolized by hydroxylation of the phenyl ring and *N*-demethylation. In addition, 3′-fluoromethcathinone-imine was detected as an unusual metabolite.

5.3.2 3′,4′-Methylenedioxy-N-alkylated Cathinones (3′,4′-Methylenedioxycathinones, bk-MDAs)

5.3.2.1 3′,4′-Methylenedioxymethcathinone (bk-MDMA, Methylone)

Kamata et al. (2006) reported on the 3′,4′-methylenedioxymethcathinone (methylone) metabolism in human and rat. According to their report, while a significant portion of methylone is excreted in urine as an unchanged form, methylone is mainly metabolized via *N*-demethylation and demethylenation, followed by *O*-methylation pathways (Fig. 5.14); reduction of the β-ketone group was not observed. Resultant methylenedioxy ring-opening metabolites, 3′-hydroxy-4′-methoxymethcathinone, and 4′-hydroxy-3′-methoxymethcathinone, are mostly excreted in urine in their conjugated forms. Based on the results from both human and in vivo study using rat (5 mg/kg i.p.), the most abundant urinary metabolite is

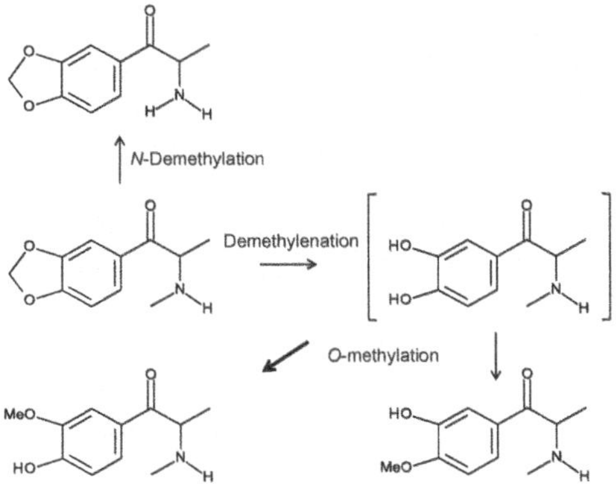

Fig. 5.14 Major phase I metabolic pathways for 3′,4′-methylenedioxymethcathinone (adapted from Kamata et al. 2006). Bold arrow indicates the more abundant pathway

4′-hydroxy-3′-methoxymethcathinone. *N*-Demethyl-metabolite is partially conjugated with dicarboxylic acid. 3′-Hydroxy-4′-methoxycathinone and 4′-hydroxy-3′-methoxycathinone were also detected in rat urine (dose: 20 mg/kg s.c.) (Židková et al. 2017).

5.3.2.2 2-Methylamino-1-(3′,4′-methylenedioxyphenyl)butan-1-one (bk-MBDB, Butylone) and 2-Ethylamino-1-(3,4-methylenedioxyphenyl)propan-1-one (bk-MDEA, Ethylone)

The metabolism of 2-methylamino-1-(3′,4′-methylenedioxyphenyl)butan-1-one (bk-MBDB) and 2-ethylamino-1-(3,4-methylenedioxyphenyl)propan-1-one (bk-MDEA) in human was studied by Zaitsu et al. (2009). In their report, both drugs were found to be excreted as their unchanged forms in urine similar to methylone, and some of these unchanged drugs were also conjugated. For the metabolites, the major metabolic pathways were common for bk-MBDB and bk-MDEA: reduction of the β-ketone moiety, *N*-dealkylation, and demethylenation followed by *O*-methylation (Fig. 5.15). The most abundant metabolites are 4′-hydroxy-3′-methoxy metabolites for both bk-MBDB and bk-MDEA. Most of 4′-hydroxy-3′-methoxy and 3′-hydroxy-4′-methoxy metabolites are excreted as their conjugated forms, while some part of β-ketone-reduced hydroxyl-metabolite are also conjugated.

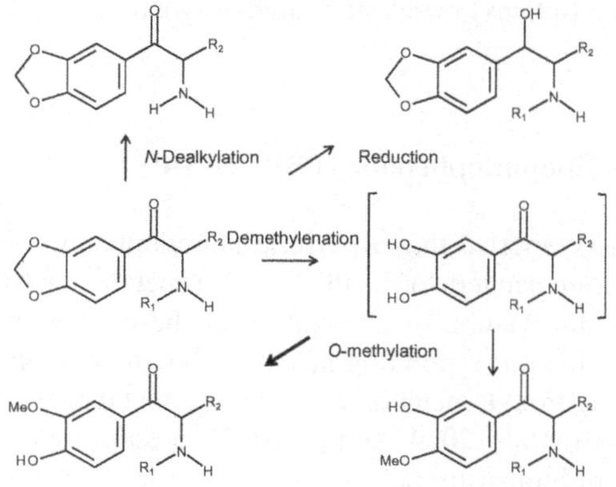

Fig. 5.15 Major phase I metabolic pathways for 2-methylamino-1-(3′,4′-methylenedioxyphenyl)-butan-1-one (bk-MBDB) and 2-ethylamino-1-(3,4-methylenedioxyphenyl)propan-1-one (bk-MDEA) (adapted from Zaitsu et al. 2009). bk-MBDB: R1 = Me, R2 = Et, bk-MDEA: R1 = Et, R2 = Me. Bold arrow indicates the more abundant pathway

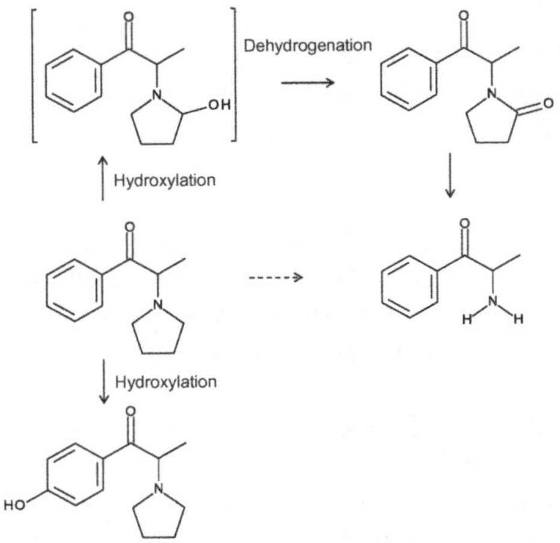

Fig. 5.16 Major phase I metabolic pathways for α-pyrrolidinopropiophenone (adapted from Springer et al. 2003c)

5.3.3 *N-Pyrrolidine Cathinone with/Without Ring Substituents*

5.3.3.1 α-Pyrrolidinopropiophenone (PPP, α-PPP)

It was demonstrated that α-pyrrolidinopropiophenone (PPP) is extensively metabolized in rats by 4′-hydroxylation of the phenyl ring, hydroxylation of 2″-position of the pyrrolidine ring followed by dehydrogenation to the corresponding lactam and degradation of the pyrrolidine ring to the primary amine (Springer et al. 2003c) (Fig. 5.16). Hydroxyl-phenyl metabolites are partly excreted as glucuronides and/or sulfates.

5.3.3.2 α-Pyrrolidinobutiophenone (PBP, α-PBP)

Matsuta et al. (2015) reported that α-pyrrolidinobutiophenone (PBP) is excreted in human urine as its unchanged form. PBP is also metabolized by reduction of the β-ketone group, hydroxylation of 2″-position of the pyrrolidine ring followed by dehydrogenation to the corresponding lactam (2″-oxo-metabolite) (Fig. 5.17). The β-ketone-reduced hydroxyl metabolite is further metabolized to the conjugated glucuronide. Namera et al. (2014) confirmed the presence of unchanged PBP and 2″-oxo-metabolite in human urine.

5.3.3.3 α-Pyrrolidinovalerophenone (PVP, α-PVP)

Tyrkkö et al. (2013) reported that PVP is extensively metabolized in human by reduction of the β-ketone group, hydroxylation followed by dehydrogenation of the

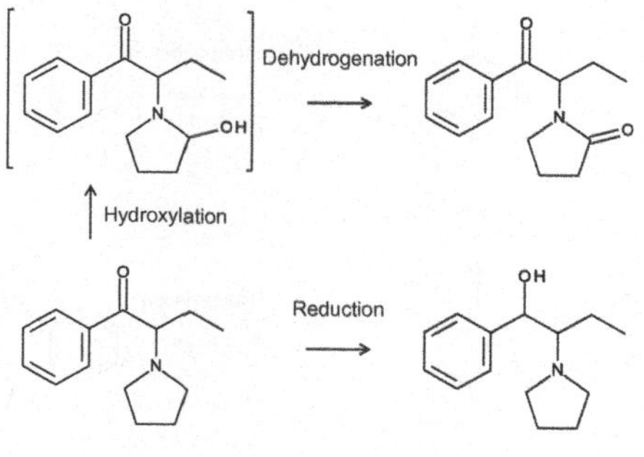

Fig. 5.17 Major phase I metabolic pathways for α-pyrrolidinobutiophenone (adapted from Matsuta et al. 2015)

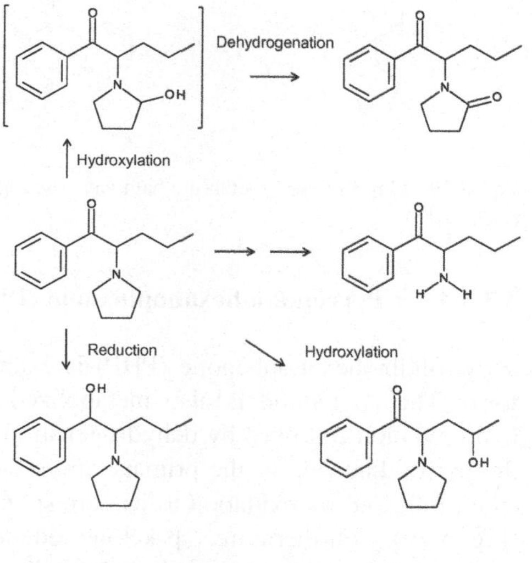

Fig. 5.18 Major phase I metabolic pathways for α-pyrrolidinovalerophenone (adapted from Tyrkkö et al. 2013, Shima et al. 2014 and Negreira et al. 2015)

pyrrolidine ring and degradation of the pyrrolidine ring to the primary amine, hydroxylation of the side chain, and some combined forms (Fig. 5.18). Some of the abovementioned metabolic pathways are also confirmed in 19 human urine samples (Shima et al. 2014). However, the pyrrolidine ring-opening pathway to the corresponding primary amine has not been confirmed in human. It was suggested that β-ketone-reduced hydroxyl metabolites are more abundant than 2″-oxo metabolites, and *O*-glucuronides of β-ketone-reduced hydroxyl metabolites were directly detected in human urine (Shima et al. 2014). In addition to the abovementioned pathways, Sauer et al. (2009) confirmed hydroxylation of the phenyl ring in rats. Phase II metabolism by glucuronidation and/or sulfation of these hydroxyl metabolites is also expected in rats. PVP metabolism by human liver microsomes and human liver cytosol was described by Negreira et al. (2015), where the abovementioned pathways were also confirmed; they also confirmed the presence of some glucuronides but not sulfates for PVP.

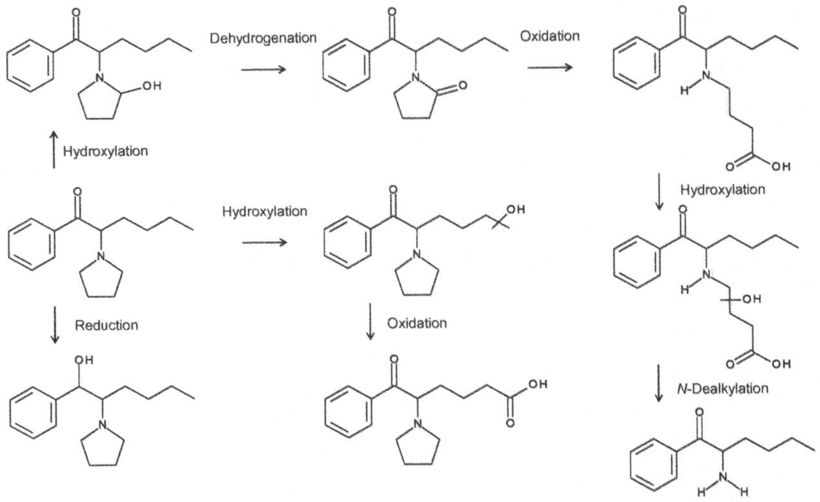

Fig. 5.19 Major phase I metabolic pathways for α-pyrrolidinohexanophenone (adapted from Paul et al. 2015)

5.3.3.4 α-Pyrrolidinohexanophenone (PHP, α-PHP, PV-7)

α-Pyrrolidinohexanophenone (PHP) is excreted in human urine as an unchanged form. The compound is also metabolized by reduction of the β-ketone group, hydroxylation followed by dehydrogenation of the pyrrolidine ring, degradation of the pyrrolidine ring to the primary amine, and hydroxylation of the aliphatic side chain followed by oxidation to its corresponding carboxylic acid (Paul et al. 2015) (Fig. 5.19). Furthermore, β-ketone-reduced hydroxyl metabolite, side-chain hydroxylated metabolite, and carboxylic acid are conjugated to glucuronides. Hisatsune et al. (2015) confirmed the abovementioned metabolic pathways in human urine: reduction of the β-ketone group, hydroxylation followed by dehydrogenation of the pyrrolidine ring, and hydroxylation of the aliphatic side chain followed by oxidation to its corresponding carboxylic acid.

5.3.3.5 α-Pyrrolidinoheptanophenone (PV-8)

Swortwood et al. (2016b) reported the metabolic profiles of α-pyrrolidinoheptanophenone (PV-8) in human hepatocytes and urine, and elucidated the metabolic pathways as follows: *N*-dealkylation, reduction of the β-ketone group, hydroxylation of side chain followed by oxidation to the corresponding carboxylic acid, hydroxylation of 2″-position of the pyrrolidine ring followed by dehydrogenation to

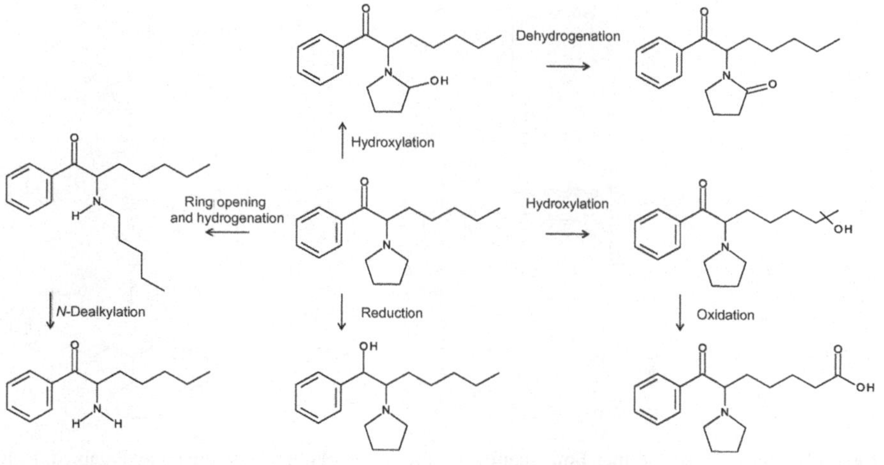

Fig. 5.20 Major phase I metabolic pathways for α-pyrrolidinoheptanophenone (adapted from Swortwood et al. 2016b)

the corresponding lactam and degradation of the pyrrolidine ring to the primary amine (Fig. 5.20). In human urine, aliphatic hydroxylation seems to be the most abundant among the identified pathways. This is due to the extension of the aliphatic side chain.

5.3.3.6 α-Pyrrolidinooctanophenone (POP, α-POP, PV9)

The metabolic profiles of α-pyrrolidinooctanophenone (POP) in human urine were described by Shima et al. (2015). They elucidated major metabolic pathways as follows: reduction of the β-ketone group, hydroxylation of the side chain followed by oxidation to the corresponding carboxylic acid, hydroxylation of the aliphatic side chain followed by oxidation to the corresponding ketone, and hydroxylation of 2″-position of the pyrrolidine ring followed by dehydrogenation to the corresponding lactam (Fig. 5.21). The major metabolic pathways for POP seem to be different from those of α-PVP and α-pyrrolidinobutiophenone (PBP), likely due to the extension of the aliphatic side chain as mentioned above. Glucuronides for the β-ketone-reduced hydroxyl and carboxylic acid metabolites were also detected.

5.3.3.7 4′-Methyl-α-pyrrolidinopropiophenone (MPPP)

4′-Methyl-α-pyrrolidinopropiophenone (MPPP) is metabolized by hydroxylation of the 4′-methyl moiety followed by oxidation to the corresponding carboxylic acid, hydroxylation of 2″-position of the pyrrolidine ring followed by dehydrogenation to the corresponding lactam, hydroxylation of the α-carbon followed by deamination,

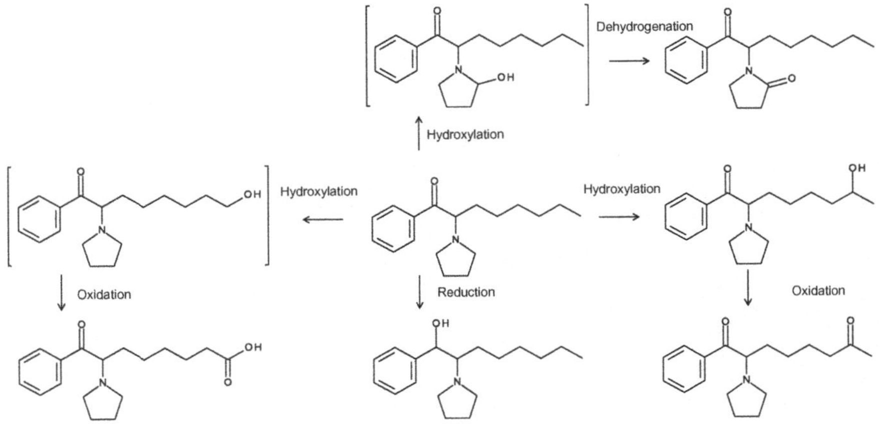

Fig. 5.21 Major phase I metabolic pathways for α-pyrrolidinooctanophenone (adapted from Shima et al. 2015)

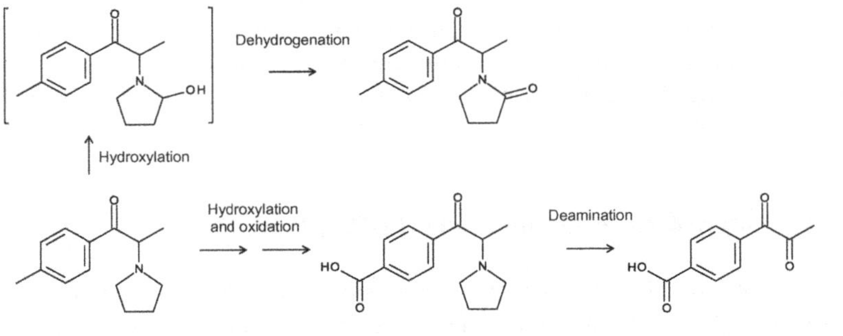

Fig. 5.22 Major phase I metabolic pathways for 4′-methyl-α-pyrrolidinopropiophenone (adapted from Springer et al. 2003b)

and further biotransformation to the corresponding benzoic acid in rat (Springer et al. 2003b) (Fig. 5.22). However, reduction of the β-ketone group is not observed for MPPP. This phenomenon is similar to that observed in metabolism of methylone as mentioned before. Also, the carboxylic acid metabolites are excreted as their acyl glucuronides.

5.3.3.8 4′-Methyl-α-pyrrolidinobutiophenone (MPBP)

The metabolism of 4′-methyl-α-pyrrolidinobutiophenone (MPBP) in rat was described by Peters et al. (2005). The elucidated major metabolic pathways are as follows: reduction of the β-ketone group, hydroxylation of the 4′-methyl moiety followed by further oxidation to the corresponding carboxylic acid, hydroxylation of 2″-position

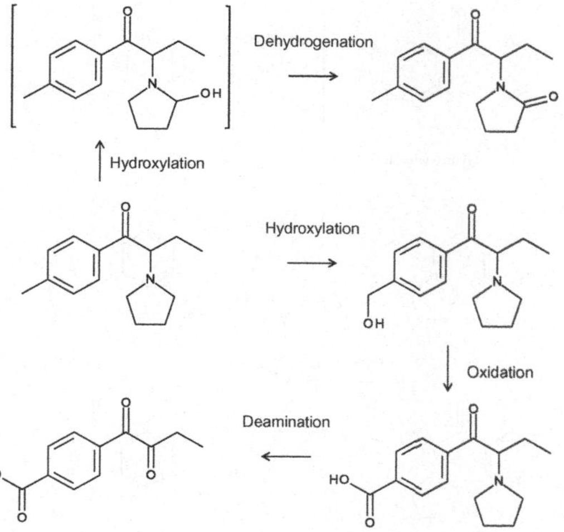

Fig. 5.23 Major phase I metabolic pathways for 4′-methyl-α-pyrrolidinobutiophenone (adapted from Peters et al. 2005)

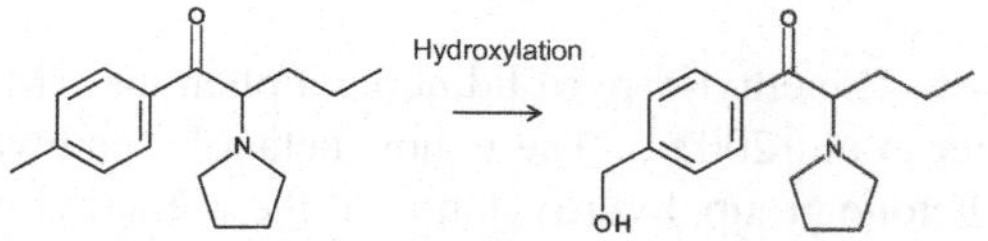

Fig. 5.24 Major phase I metabolic pathway of 4′-methyl-α-pyrrolidinovalerophenone

of the pyrrolidine ring followed by dehydrogenation to the corresponding lactam, and hydroxylation of the α-carbon followed by deamination (Fig. 5.23). Carboxylic acid and hydroxyl metabolites are partly excreted as glucuronides and/or sulfates.

5.3.3.9 4′-Methyl-α-pyrrolidinovalerophenone (MPVP, Pyrovalerone, O-2371)

The hydroxylation of the 4′-methyl group is the only metabolic pathway for 4′-methyl-α-pyrrolidinovalerophenone (pyrovalerone) in rat (Fig. 5.24), though this hydroxyl metabolite is not detected in human, where only the unchanged pyrovalerone is identified (Michaelis et al. 1970). This metabolic pathway was confirmed by Lho et al. (1996) and Shin et al. (1996). However, no further detailed information on pyrovalerone metabolism is available.

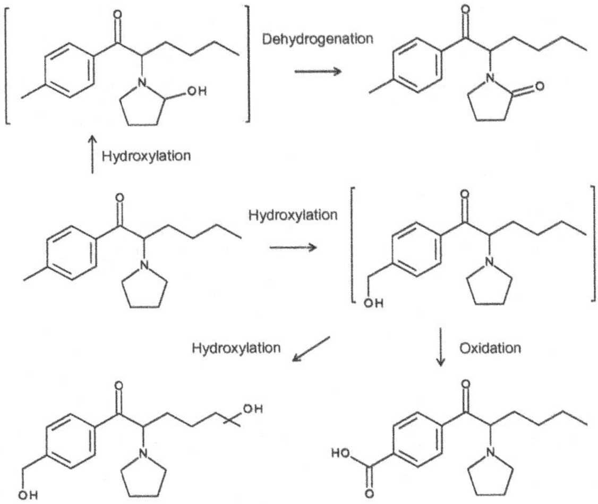

Fig. 5.25 Major phase I metabolic pathways for 4′-methyl-α-pyrrolidinohexiophenone (adapted from Springer et al. 2003d)

5.3.3.10 4′-Methyl-α-pyrrolidinohexiophenone (MPHP)

The metabolism of 4′-methyl-α-pyrrolidinohexiophenone (MPHP) in rat was reported by Springer et al. (2003d). The major metabolic pathways are as follows: reduction of the β-ketone group, hydroxylation of the 4′-methyl moiety followed by further oxidation to the corresponding carboxylic acid, hydroxylation of 2″-position of the pyrrolidine ring followed by dehydrogenation to the corresponding lactam, and hydroxylation of the aliphatic side chain (Fig. 5.25). In contrast to MPPP and MPBP, deamination pathway is not observed in MPHP metabolism. Carboxylic and hydroxyl metabolites are partly excreted as glucuronides and/or sulfates.

5.3.3.11 4′-Methoxy-α-pyrrolidinopropiophenone (MOPPP)

4′-Methoxy-α-pyrrolidinopropiophenone (MOPPP) metabolism in rat was described by Springer et al. (2003b) (Fig. 5.26). By analogy to other 4-methoxyamphetamines, *O*-demethylation of the 4′-methoxyl group is considered as one of the main metabolic pathways. Hydroxylation of 2″-position of the pyrrolidine ring followed by dehydrogenation to the corresponding lactam, and hydroxylation of the α-carbon followed by deamination are also elucidated. In addition, hydroxyl-methoxy metabolites, which are generated by hydroxylation at the 3′-position of 4′-hydroxy metabolites followed by the *O*-methylation pathway, are detected. However, as with MPPP, reduction of the β-ketone group is not confirmed. The abovementioned hydroxyl metabolites and hydroxyl-methoxy metabolites are partly excreted as glucuronides and/or sulfates (Springer et al. 2003b).

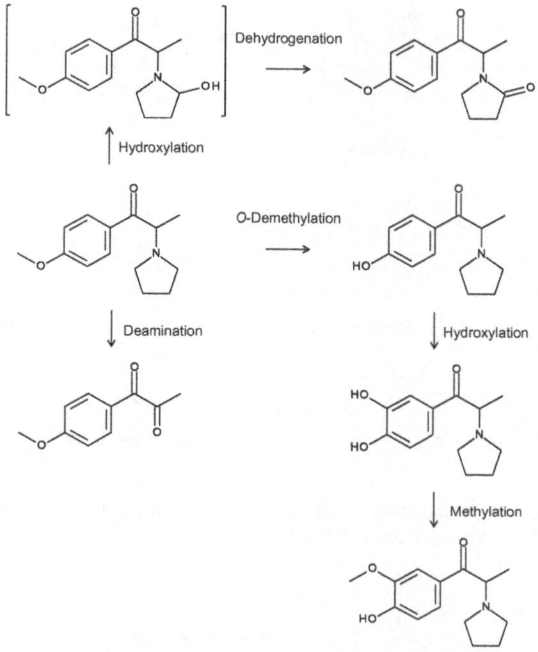

Fig. 5.26 Major phase I metabolic pathways for 4′-methoxy-α-pyrrolidinopropiophenone (adapted from Springer et al. 2003b)

*5.3.4 3′,4′-Methylenedioxy-*N*-pyrrolidine Cathinone Derivatives*

5.3.4.1 3′,4′-Methylenedioxy-α-pyrrolidinopropiophenone (MDPPP)

3′,4′-Methylenedioxy-α-pyrrolidinopropiophenone (MDPPP) metabolism in rat was described by Springer et al. (2003a) (Fig. 5.27). MDPPP is mainly metabolized by demethylenation followed by *O*-methylation, hydroxylation of 2″-position of the pyrrolidine ring followed by dehydrogenation to the corresponding lactam and hydroxylation of the α-carbon followed by deamination. Reduction of the β-ketone group and hydroxylation of the aliphatic side chain are not confirmed. These phenomena are similar to those for methylone and MPPP. Demethylenated metabolites are excreted as glucuronides and/or sulfates.

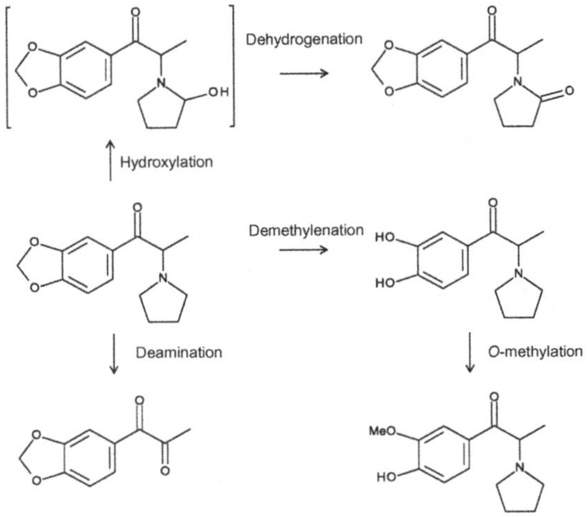

Fig. 5.27 Major phase I metabolic pathways for 3′,4′-methylenedioxy-α-pyrrolidinopropiophenone (adapted from Springer et al. 2003a)

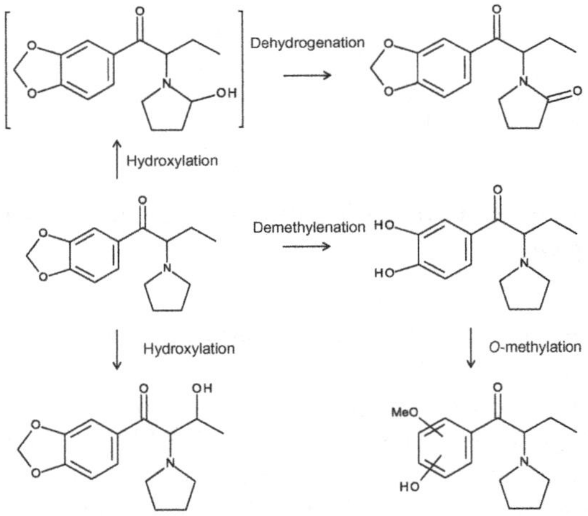

Fig. 5.28 Major phase I metabolic pathways for 3′,4′-Methylenedioxy-α-pyrrolidinobutiophenone (adapted from Meyer et al. 2014)

5.3.4.2 3′,4′-Methylenedioxy-α-pyrrolidinobutiophenone (MDPBP)

Meyer et al. (2014) reported the in vivo and in vitro metabolism of 3′,4′-methylenedioxy-α-pyrrolidinobutiophenone (MDPBP) (Fig. 5.28), elucidating demethylenation followed by *O*-methylation, hydroxylation of the aliphatic side chain and hydroxylation of 2″-position of the pyrrolidine ring followed by dehydrogenation to the corresponding lactam. The demethylenated metabolites are partly excreted as glucuronides and/or sulfates.

5.3.4.3 3′,4′-Methylenedioxy-α-pyrrolidinovalerophenone (MDPVP, MDPV)

Meyer et al. (2010a) described the metabolism of 3′,4′-methylenedioxy-α-pyrrolidinovalerophenone (MDPV) in rat and human, and Negreira et al. (2015) reported the in vitro metabolism of MDPV by human liver microsomes and human liver cytosol (Fig. 5.29). As with MDPPP and bk-MDAs, MDPV is mainly metabolized by reduction of the β-ketone group, demethylenation followed by *O*-methylation and hydroxylation of 2″-position of the pyrrolidine ring, followed by dehydrogenation to the corresponding lactam. In vitro phase II metabolism study demonstrated that demethylenated metabolites are conjugated to glucuronides and/or sulfates (Negreira et al. 2015).

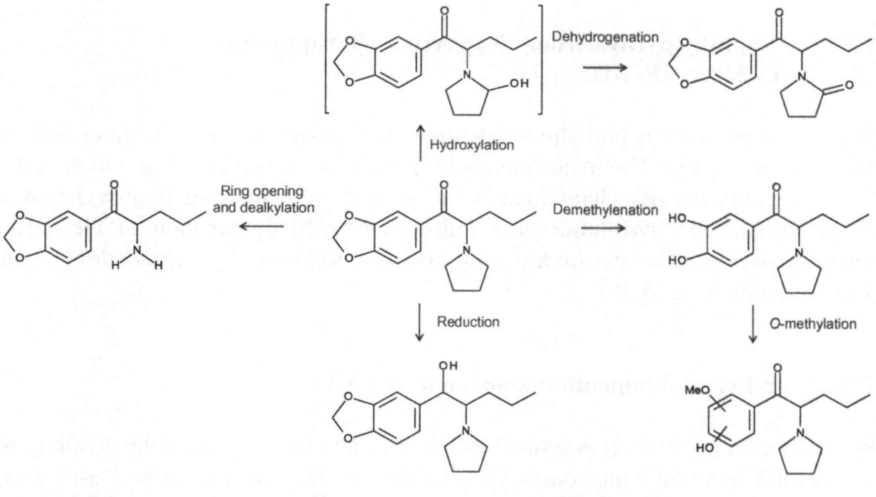

Fig. 5.29 Major phase I metabolic pathways for 3′,4′-methylenedioxy-α-pyrrolidinovalerophenone (adapted from Meyer et al. 2010a and Negreira et al. 2015)

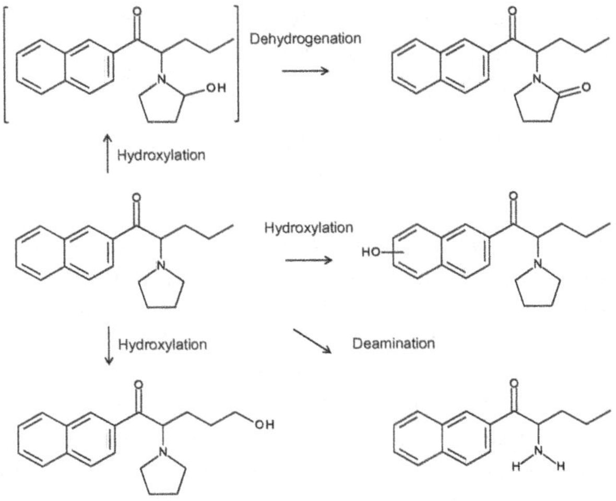

Fig. 5.30 Major phase I metabolic pathways for naphthylpyrovalerone (adapted from Meyer et al. 2013)

5.3.5 *Other* N-*pyrrolidine Cathinone Derivatives*

5.3.5.1 Naphthylpyrovalerone (Naphyrone, β-Naphyrone, O-2482, NRG-1)

The metabolism of naphthylpyrovalerone (naphyrone) in rat was described by Meyer et al. (2013). The major metabolic pathways for naphyrone are hydroxylation of the aliphatic side chain, hydroxylation of the naphthyl ring, hydroxylation of 2″-position of the pyrrolidine ring followed by dehydrogenation to the corresponding lactam, and pyrrolidine ring opening followed by degradation to the primary amine (Fig. 5.30).

5.3.5.2 α-Pyrrolidinopentiothiophenone (α-PVT)

Swortwood et al. (2016a) described the in vitro and in vivo metabolic profiling of α-pyrrolidinopentiothiophenone (α-PVT) (Fig. 5.31). In rat urine, α-PVT is excreted as an unchanged form. They also elucidated the major metabolic pathways as follows: reduction of the β-ketone group, hydroxylation of the thiophene ring, hydroxylation of the aliphatic side chain, and hydroxylation of 2″-position of the pyrrolidine ring followed by dehydrogenation to the corresponding lactam. It should be noted that the hydroxyl-thiophene metabolite shows keto-enol tautomerism; therefore, hydrolylated-thiophene metabolite is detectable as its ketone form. The hydrolylated-pyrrolidine metabolite is further excreted as glucuronide.

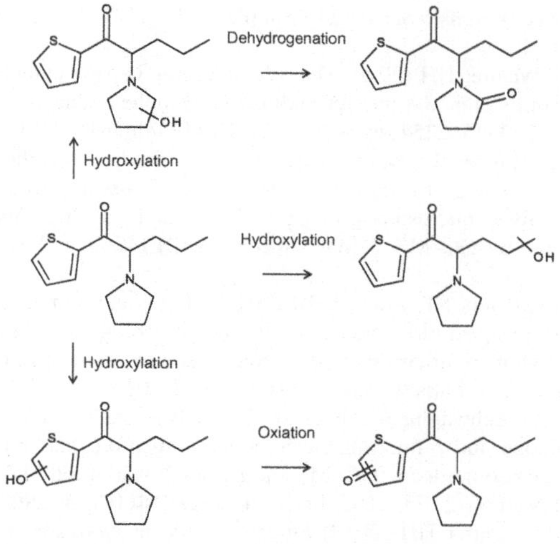

Fig. 5.31 Major phase I metabolic pathways for α-pyrrolidinopentiothiophenone (adapted from Swortwood et al. 2016a)

References

Brenneisen R, Geisshüsler S, Schorno X (1986) Metabolism of cathinone to (−)-norephedrine and (−)-norpseudoephedrine. J Pharm Pharmacol 38(4):298–300. https://doi.org/10.1111/j.2042-7158.1986.tb04571.x

Caldwell J, Dring LG, Williams RT (1972) Metabolism of [^{14}C]methamphetamine in man, the guinea pig and the rat. Biochem J 129(1):11–22. https://doi.org/10.1042/bj1290011

Guantai AN, Maitai CK (1983) Metabolism of cathinone to d-norpseudoephedrine in humans. J Pharm Sci 72(10):1217–1218. https://doi.org/10.1002/jps.2600721029

Hisatsune K, Zaitsu K, Kusano M et al (2015) Determination of newly encountered designer drugs α-PHP and acetylfentanyl in an acute intoxication case by LC/Q-TOFMS. In: Proceedings of the 53rd annual meeting of the international association of forensic toxicologists, Fiorentina, Italy, 31 August 2015

Kamata HT, Shima N, Zaitsu K et al (2006) Metabolism of the recently encountered designer drug, methylone, in humans and rats. Xenobiotica 36(8):709–723. https://doi.org/10.1080/00498250600780191

Kelly JP (2011) Cathinone derivatives: a review of their chemistry, pharmacology and toxicology. Drug Test Anal 3(7–8):439–453. https://doi.org/10.1002/dta.313

Khreit OIG, Grant MH, Zhang T et al (2013) Elucidation of the phase I and phase II metabolic pathways of (±)-4′-methylmethcathinone (4-MMC) and (±)-4′-(trifluoromethyl)methcathinone (4-TFMMC) in rat liver hepatocytes using LC–MS and LC–MS2. J Pharm Biomed Anal 72 (Supplement C):177–185. https://doi.org/10.1016/j.jpba.2012.08.015

Lho D-S, Lee J, Kim S et al (1996) Identification of a pyrovalerone metabolite in the rat by gas chromatography-mass spectrometry and determination of pyrovalerone by gas chromatography-nitrogen-phosphorus detection. J Chromatogr B Biomed Appl 687(1):253–259. https://doi.org/10.1016/S0378-4347(96)00183-1

Matsuta S, Shima N, Kamata H et al (2015) Metabolism of the designer drug α-pyrrolidinobutiophenone (α-PBP) in humans: identification and quantification of the phase

I metabolites in urine. Forensic Sci Int 249(Supplement C):181–188. https://doi.org/10.1016/j.forsciint.2015.02.004

Meyer MR, Peters FT, Maurer HH (2008) The role of human hepatic cytochrome P450 isozymes in the metabolism of racemic 3,4-methylenedioxymethamphetamine and its enantiomers. Drug Metab Dispos 36(11):2345–2354. https://doi.org/10.1124/dmd.108.021543

Meyer MR, Du P, Schuster F, Maurer HH (2010a) Studies on the metabolism of the α-pyrrolidinophenone designer drug methylenedioxy-pyrovalerone (MDPV) in rat and human urine and human liver microsomes using GC–MS and LC–high-resolution MS and its detectability in urine by GC–MS. J Mass Spectr 45(12):1426–1442. https://doi.org/10.1002/jms.1859

Meyer MR, Wilhelm J, Peters FT, Maurer HH (2010b) Beta-keto amphetamines: studies on the metabolism of the designer drug mephedrone and toxicological detection of mephedrone, butylone, and methylone in urine using gas chromatography–mass spectrometry. Anal Bioanal Chem 397(3):1225–1233. https://doi.org/10.1007/s00216-010-3636-5

Meyer MR, Vollmar C, Schwaninger AE et al (2012) New cathinone-derived designer drugs 3-bromomethcathinone and 3-fluoromethcathinone: studies on their metabolism in rat urine and human liver microsomes using GC–MS and LC–high-resolution MS and their detectability in urine. J Mass Spectr 47(2):253–262. https://doi.org/10.1002/jms.2960

Meyer MR, Prosser D, Maurer HH (2013) Studies on the metabolism and detectability of the designer drug β-naphyrone in rat urine using GC-MS and LC-HR-MS/MS. Drug Test Anal 5(4):259–265. https://doi.org/10.1002/dta.1443

Meyer MR, Mauer S, Meyer GMJ et al (2014) The in vivo and in vitro metabolism and the detectability in urine of 3′,4′-methylenedioxy-alpha-pyrrolidinobutyrophenone (MDPBP), a new pyrrolidinophenone-type designer drug, studied by GC-MS and LC-MSn. Drug Test Anal 6(7–8):746–756. https://doi.org/10.1002/dta.1559

Michaelis W, Russel JH, Schindler O (1970) Metabolism of pyrovalerone hydrochloride. J Med Chem 13(3):497–503. https://doi.org/10.1021/jm00297a036

Namera A, Konuma K, Kawamura M et al (2014) Time-course profile of urinary excretion of intravenously administered α-pyrrolidinovalerophenone and α-pyrrolidinobutiophenone in a human. Forensic Toxicol 32(1):68–74. https://doi.org/10.1007/s11419-013-0203-8

Negreira N, Erratico C, Kosjek T et al (2015) In vitro phase I and phase II metabolism of α-pyrrolidinovalerophenone (α-PVP), methylenedioxypyrovalerone (MDPV) and methedrone by human liver microsomes and human liver cytosol. Anal Bioanal Chem 407(19):5803–5816. https://doi.org/10.1007/s00216-015-8763-6

Paul BD, Cole KA (2001) Cathinone (Khat) and methcathinone (CAT) in urine specimens: a gas chromatographic-mass spectrometric detection procedure. J Anal Toxicol 25(7):525–530. https://doi.org/10.1093/jat/25.7.525

Paul M, Bleicher S, Guber S et al (2015) Identification of phase I and II metabolites of the new designer drug α-pyrrolidinohexiophenone (α-PHP) in human urine by liquid chromatography quadrupole time-of-flight mass spectrometry (LC-QTOF-MS). J Mass Spectr 50(11):1305–1317. https://doi.org/10.1002/jms.3642

Pawlik E, Plässer G, Mahler H, Daldrup T (2012) Studies on the phase I metabolism of the new designer drug 3-fluoromethcathinone using rabbit liver slices. Int J Legal Med 126(2):231–240. https://doi.org/10.1007/s00414-011-0601-6

Pedersen AJ, Reitzel LA, Johansen SS, Linnet K (2013) In vitro metabolism studies on mephedrone and analysis of forensic cases. Drug Test Anal 5(6):430–438. https://doi.org/10.1002/dta.1369

Peters FT, Meyer MR, Fritschi G, Maurer HH (2005) Studies on the metabolism and toxicological detection of the new designer drug 4′-methyl-α-pyrrolidinobutyrophenone (MPBP) in rat urine using gas chromatography–mass spectrometry. J Chromatogr B 824(1):81–91. https://doi.org/10.1016/j.jchromb.2005.07.003

Sauer C, Peters FT, Haas C, Meyer MR, Fritschi G, Maurer HH (2009) New designer drug α-pyrrolidinovalerophenone (PVP): studies on its metabolism and toxicological detection in rat urine using gas chromatographic/mass spectrometric techniques. J Mass Spectrom 44:952–964

Schreiber EC, Min BH, Zeiger AV, Lang JF (1968) Metabolism of diethylpropion-1-C^{14} hydrochloride by the human. J Pharm Exp Ther 159(2):372–378

Shima N, Katagi M, Kamata HZ et al (2008) Urinary excretion of the main metabolites of 3,4-methylenedioxymethamphetamine (MDMA), including the sulfate and glucuronide of 4-hydroxy-3-methoxymethamphetamine (HMMA), in humans and rats. Xenobiotica 38(3): 314–324. https://doi.org/10.1080/00498250701802506

Shima N, Katagi M, Kamata H et al (2013) Urinary excretion and metabolism of the newly encountered designer drug 3,4-dimethylmethcathinone in humans. Forensic Toxicol 31(1): 101–112. https://doi.org/10.1007/s11419-012-0172-3

Shima N, Katagi M, Kamata H et al (2014) Metabolism of the newly encountered designer drug α-pyrrolidinovalerophenone in humans: identification and quantitation of urinary metabolites. Forensic Toxicol 32(1):59–67. https://doi.org/10.1007/s11419-013-0202-9

Shima N, Kakehashi H, Matsuta S et al (2015) Urinary excretion and metabolism of the α-pyrrolidinophenone designer drug 1-phenyl-2-(pyrrolidin-1-yl)octan-1-one (PV9) in humans. Forensic Toxicol 33(2):279–294. https://doi.org/10.1007/s11419-015-0274-9

Shin H-S, Shin Y-SO, Lee S, Park BB (1996) Detection and identification of pyrovalerone and its hydroxylated metabolite in the rat. J Anal Toxicol 20(7):568–572. https://doi.org/10.1093/jat/20.7.568

Springer D, Fritschi G, Maurer HH (2003a) Metabolism and toxicological detection of the new designer drug 3′,4′-methylenedioxy-α-pyrrolidinopropiophenone studied in urine using gas chromatography–mass spectrometry. J Chromatogr B 793(2):377–388. https://doi.org/10.1016/S1570-0232(03)00350-7

Springer D, Fritschi G, Maurer HH (2003b) Metabolism and toxicological detection of the new designer drug 4′-methoxy-α-pyrrolidinopropiophenone studied in rat urine using gas chromatography–mass spectrometry. J Chromatogr B 793(2):331–342. https://doi.org/10.1016/S1570-0232(03)00334-9

Springer D, Fritschi G, Maurer HH (2003c) Metabolism of the new designer drug α-pyrrolidinopropiophenone (PPP) and the toxicological detection of PPP and 4′-methyl-α-pyrrolidinopropiophenone (MPPP) studied in rat urine using gas chromatography-mass spectrometry. J Chromatogr B 796(2):253–266. https://doi.org/10.1016/j.jchromb.2003.07.008

Springer D, Peters FT, Fritschi G, Maurer HH (2003d) New designer drug 4′-methyl-α-pyrrolidinohexanophenone: studies on its metabolism and toxicological detection in urine using gas chromatography–mass spectrometry. J Chromatogr B 789(1):79–91. https://doi.org/10.1016/S1570-0232(03)00043-6

Staack RF, Maurer HH (2005) Metabolism of designer drugs of abuse. Curr Drug Metab 6(3):259–274. https://doi.org/10.2174/1389200054021825

Staack RF, Theobald DS, Paul LD et al (2004) Identification of human cytochrome P450 2D6 as major enzyme involve in the *O*-demethylation of the designer drug *p*-methoxymethamphetamine. Drug Metab Dispos 2(4):379–381. https://doi.org/10.1124/dmd.32.4.379

Swortwood MWA, Concheiro M, Marilyn AH (2016a) In vitro, in vivo and in silico metabolic profiling of α-pyrrolidinopentiothiophenone, a novel thiophene stimulant. Bioanalysis 8(1):65–82. https://doi.org/10.4155/bio.15.237

Swortwood MJ, Ellefsen KN, Wohlfarth A et al (2016b) First metabolic profile of PV8, a novel synthetic cathinone, in human hepatocytes and urine by high-resolution mass spectrometry. Anal Bioanal Chem 408(18):4845–4856. https://doi.org/10.1007/s00216-016-9599-4

Testa B, Beckett AH (1972) Studies on the metabolism of diethylpropion: I. Analytical procedure. J Chromatogr A 71(1):39–54. https://doi.org/10.1016/S0021-9673(01)85688-0

Testa B, Beckett AH (1973) Metabolism and excretion of diethylpropion in man under acidic urine conditions. J Pharm Pharmacol 25(2):119–124. https://doi.org/10.1111/j.2042-7158.1973.tb10604.x

Tyrkkö E, Pelander A, Ketola RA, Ojanperä I (2013) In silico and in vitro metabolism studies support identification of designer drugs in human urine by liquid chromatography/quadrupole-time-of-flight mass spectrometry. Anal Bioanal Chem 405(21):6697–6709. https://doi.org/10.1007/s00216-013-7137-1

Valente MJ, Guedes de Pinho P, de Lourdes Bastos M et al (2014) Khat and synthetic cathinones: a review. Arch Toxicol 88(1):15–45. https://doi.org/10.1007/s00204-013-1163-9

Wikström M, Thelander G, Nyström I, Kronstrand R (2010) Two fatal intoxications with the new designer drug methedrone (4-methoxymethcathinone). J Anal Toxicol 34(9):594–598. https://doi.org/10.1093/jat/34.9.594

Zaitsu K, Katagi M, Kamata T et al (2008) Determination of a newly encountered designer drug "p-methoxyethylamphetamine" and its metabolites in human urine and blood. Forensic Sci Int 177(1):77–84. https://doi.org/10.1016/j.forsciint.2007.11.001

Zaitsu K, Katag M, Kamata HT et al (2009) Determination of the metabolites of the new designer drugs bk-MBDB and bk-MDEA in human urine. Forensic Sci Int 188(1):131–139. https://doi.org/10.1016/j.forsciint.2009.04.001

Zaitsu K, Katagi M, Tatsuno M et al (2011) Recently abused β-keto derivatives of 3,4-methylenedioxyphenylalkylamines: a review of their metabolisms and toxicological analysis. Forensic Toxicol 29(2):73–84. https://doi.org/10.1007/s11419-011-0111-8

Zaitsu K, Katagi M, Tsuchihashi H, Ishii A (2014) Recently abused synthetic cathinones, α-pyrrolidinophenone derivatives: a review of their pharmacology, acute toxicity, and metabolism. Forensic Toxicol 32(1):1–8. https://doi.org/10.1007/s11419-013-0218-1

Zawilska JB, Wojcieszak J (2013) Designer cathinones—an emerging class of novel recreational drugs. Forensic Sci Int 231(1):42–53. https://doi.org/10.1016/j.forsciint.2013.04.015

Žídková M, Linhart I, Balíková M et al (2017) Identification of three new phase II metabolites of a designer drug methylone formed in rats by *N*-demethylation followed by conjugation with dicarboxylic acids. Xenobiotica 1–8. https://doi.org/10.1080/00498254.2017.1349964

Chapter 6
Monoamine Transporter and Receptor Interaction Profiles of Synthetic Cathinones

Linda D. Simmler

Abstract Cathinones are amphetamine-type substances with a β-keto substitution, typically emerging as drugs of abuse among novel psychoactive substances (NPS). In vitro screenings have been useful in characterizing pharmacological profiles of cathinone NPS and in obtaining fast estimates of mechanisms of action, potencies, and potential toxicity. Pharmacological targets of cathinone psychostimulants are dopamine, serotonin, and noradrenaline re-uptake transporters (DAT, SERT, and NET, respectively), as well as monoamine receptors. However, there are considerable differences among cathinones with respect to potencies and relative activity at monoaminergic targets. Foremost, cathinones act on NET, DAT, and/or SERT as inhibitors, usually also inducing reverse transport of monoamines. For many cathinones, it is difficult to predict pharmacological profiles, and consequently to anticipate acute effects, doses, and toxicity, based solely on their chemical structures. In vitro screening and pharmacological classification are therefore essential in providing translationally valuable information promptly after new cathinone derivatives emerge among drug users. This chapter provides an overview of the pharmacological profiles of NPS cathinones, which have been characterized in in vitro screenings.

Keywords Affinity · Cathinone · DAT · Dopamine · NET · Noradrenaline Monoamines · Pharmacological profiles · Release · Serotonin · SERT Synthetic cathinones · Uptake

L. D. Simmler (✉)
Department of Basic Neurosciences, University of Geneva,
Rue Michel-Servet 1, 1206 Geneva, Switzerland
e-mail: linda.simmler@unige.ch

J. B. Zawilska (ed.), *Synthetic Cathinones*, Current Topics in Neurotoxicity 12,
https://doi.org/10.1007/978-3-319-78707-7_6

6.1 Introduction

Derivatives of cathinone, referred to here as synthetic cathinones, have become popular drugs of abuse in recent years. Since 2005, 118 different synthetic cathinone molecules were reported to the European Monitoring Centre for Drugs and Drug Addiction (EMCDDA 2016). The β-keto substitution at the carbon side chain is the characteristic structural commonality of all synthetic cathinones. However, structural variations at several derivatization sites produce distinct molecules, accounting for the numerous designer cathinones on the illicit drug market. As is well known from drug discovery projects, derivatives, although similar in chemical structure, can vary considerably in potency and selectivity to target proteins (Eshleman et al. 2017). Synthetic cathinones have been designed to resemble psychostimulants such as cathinone, which is the main psychostimulant active component in *Catha edulis*. Due to the structural similarity between synthetic cathinones and cathinone or other psychostimulants, monoamine transporters and receptors in the brain are the likely targets of synthetic cathinones (Fig. 6.1). The monoamine systems, i.e., the serotonin (5-HT), dopamine (DA), and noradrenaline (NA) system, play different roles in acute intoxication with respect to both psychotropic and adverse effects. Subtle differences in chemical derivatization of cathinones can result in different relative activities at the 5-HT-, DA-, and NA systems. However, most of the over 100 synthetic cathinone derivatives have not been tested in a laboratory setting before entering the illicit drug market. Drug users are in the dark regarding dose and expected intoxication, but nevertheless they experiment by abusing these drugs, often sharing their experience regarding dose and subjective effects on Internet forums.

In vitro pharmacological testing serves as a quick assessment of the dangers of new psychoactive substances (NPS). Particularly helpful are gross assessments of which monoamine systems the substances preferentially act on, and of the risk for overdose posed by extremely potent substances. Since stimulation of the different monoamine systems results in specific intoxication, it is worth knowing which targets the new compounds act on and which types of intoxication they might induce. Typically, serotonergic effects are associated with hyperthermia, noradrenergic effects with sympathomimetic cardiostimulation, and dopaminergic effects with the reinforcing properties of a drug (Liechti 2015). The connection between in vitro pharmacology and expected toxicology or acute subjective effects is based on years of research with psychostimulants such as 3,4-methylenedioxymethamphetamine (MDMA), methamphetamine, and cocaine. Of highest interest here are the aspects of in vitro data that can be translated to clinically valuable information. Combined efforts of preclinical and clinical experts in the field have elucidated the pharmacology and toxicology of synthetic cathinones almost concurrently with the emergence of their abuse. In this chapter, I will discuss the value and limitations of in vitro pharmacological profiling and how in vitro results can be translated into predictive clinical effects.

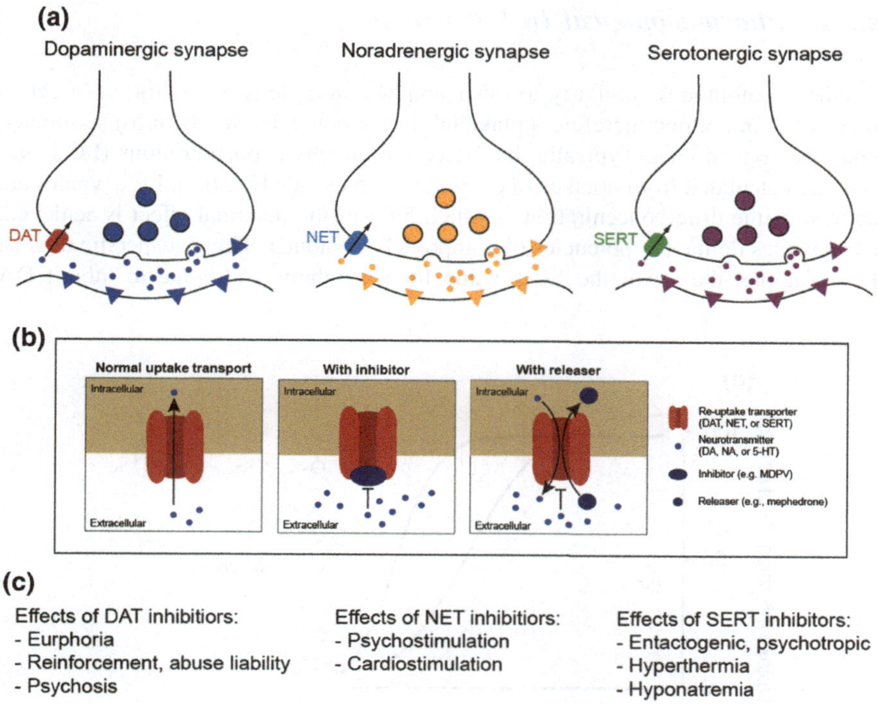

Fig. 6.1 Synaptic targets of synthetic cathinones and their typical effects. **a** Similar to amphetamine derivatives, the main pharmacological targets of synthetic cathinones are the dopamine—(DA), noradrenaline—(NA), and serotonin—(5-HT) reuptake transporters (DAT, NET, and SERT, respectively). These monoamine transporters are responsible for clearing neurotransmitters from the synaptic cleft, thereby regulating or terminating the signal transduction. **b** Normal uptake transport of neurotransmitter (left) is inhibited in the presence of an inhibitor (middle). In the presence of a releaser (right), the releaser acts as transporter substrate, preventing neurotransmitter uptake from the synaptic cleft and allowing reverse transport (Blakely et al. 2005). **c** Typical clinical subjective effects and toxicity arising from action at the respective monoamine transporters

6.2 Pharmacological Profiling of New Synthetic Cathinones In Vitro

Classical pharmacological profiling of potential targets of synthetic cathinones has proven to be a good strategy for initially understanding a new psychoactive substance. The main principles of in vitro pharmacological data and methods are described here for ease in following the further discussion on pharmacological profiles.

6.2.1 Pharmacological In Vitro Data

Synthetic cathinones, similarly to other amphetamine derivatives, foremost act as transporter inhibitors; therefore, uptake inhibition potencies are the most significant pharmacological data. Typically, half-maximal effective concentrations (EC_{50} values) are calculated from uptake concentration curves (see Fig. 6.2). EC_{50} values are defined as the drug concentration at which 50% of the maximal effect is achieved. EC_{50} values define the potencies of compounds to induce effects at specific targets. For example, the lower the EC_{50} value for a synthetic cathinone to inhibit DA

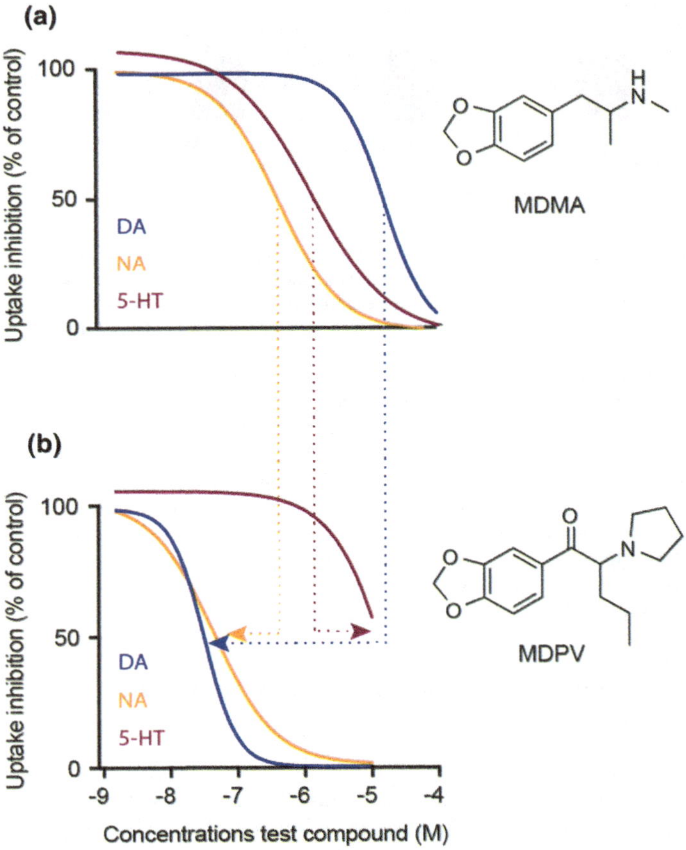

Fig. 6.2 Uptake concentration curves for 3,4-methylenedioxymethamphetamine (MDMA) and 3,4-methylenedioxypyrovalerone (MDPV). The structural differences between MDMA **a** and MDPV **b** result in significantly different uptake inhibition profiles. MDPV is much more potent at inhibiting DAT and NET than MDMA, illustrated by the leftward shifts in the dose–response curves. In contrast, MDPV is significantly less potent at inhibiting SERT than MDMA, illustrated by a rightward shift in the dose–response curve

uptake, the more potently the compound blocks the DAT. EC_{50} values are also determined for a measure of the potency of a drug to induce transporter-mediated release, using release concentration curves. To measure affinity, K_i values are calculated from the concentration at which half of the reference ligand is displaced.

6.2.2 *In Vitro Assays*

For pharmacological profiling in vitro, the target proteins of interest are typically expressed in cell cultures (see Simmler and Liechti 2017 for more details). High selectivity is assured because only one target protein is expressed per transfected cell line. Alternatively, ex vivo preparations from animal brains can be used, with the disadvantages of using animals, and the need for pharmacological blockade of other possible target proteins present in the tissue preparation. For example, selective pharmacological blockade of DAT is necessary when testing NA uptake inhibition in synaptosomes (Rothman et al. 2001). Using radioactively labeled endogenous substrates (e.g., ^{3}H-5-HT for SERT) in cultured cells or tissue preparations, transporter function can be assessed in the presence of various concentrations of test and control substances. As described above, EC_{50} values are computed from effect concentration curves as shown in Fig. 6.2. Similar assays are used to assess uptake or release. For uptake, neurotransmitter transport into cells is quantified, while for release, transport out of cells that were previously loaded with radiolabeled neurotransmitters is assessed. For receptor affinity testing, displacement of a known ligand by the test substance is determined. Membrane preparations from cultured cells are often used for affinity testing since only the presence of target protein is required. In contrast, for determination of the functional activity of a receptor ligand, like agonism or antagonism, integral systems such as cultured cells are required. For example, changes in intracellular concentration of Ca^{2+} ions due to activation of G-protein coupled receptors can be measured using commercially available kits with a Ca^{2+} indicator dye.

While the determination of drug effects at the isolated target (i.e., DAT, NET, and SERT) can characterize the direct drug action at the target protein, other physiological components can also contribute significantly to the overall effect of the drug. It has been proposed that transporter-mediated, drug-induced efflux of neurotransmitter occurs through effects on the vesicular monoamine transporter 2 (VMAT2), depleting neurotransmitter from the vesicles into the cytosol (Nickell et al. 2014). Accordingly, full assessment of release would require testing the effects of a drug on the membrane transporters (SERT, DAT, and NET) and the effects of a drug at VMAT2. Alternatively, a more physiological system, such as synaptosomes or brain slices, could be used. However, reverse transport can also occur in cell lines that only express the plasma membrane transporters but not VMAT2 (Eshleman et al. 2013; Scholze et al. 2000) and in synaptosomes when VMAT2 is inhibited (Rothman et al. 2001). Given this, testing compounds for release properties in simple assays, such as cell lines, have been useful at least for qualitative assessment of drug effects.

Synthetic cathinones have stereocenters, meaning that the molecules exist in *R*- and *S*-stereoisoforms. Typically, one of the isoforms is more active than the other (Iversen et al. 2013; Rothman et al. 2001). For example, different pharmacological profiles of *R*- and *S*-mephedrone reflect significant differences in behavioral effects induced by the two different isoforms (Gregg et al. 2015). Nevertheless, in vitro testing is usually performed using the racemic mixture of the test compounds. Because powder and pills from the illicit drug market typically also come as racemic mixtures, testing the racemic mixture is a good strategy for initial pharmacological profiling of a compound.

6.3 In Vitro Screenings for Predictions of Acute Effects and Toxicity

In vitro pharmacological assays are fairly abstract models of in vivo physiology, given that the brain is very complex, with macro- and microcircuits that regulate neuronal activity very specifically. By assessing in vitro pharmacology at specific target proteins in cell lines, possible contributions to pharmacological effects from the circuit or cellular levels are undetectable. Furthermore, physiological processes relevant for neurons, such as intracellular vesicular depletion, that may be necessary for transporter-mediated neurotransmitter release in neurons are not assessed in cell line-based assays. Despite the reductionist approach in determining pharmacological profiles, the in vitro assessments, combined with interpretations that take into account the wealth of animal and human research, are highly valid. For synthetic cathinones that lack animal and human research, the interpretation of in vitro pharmacological profiles can provide a good approximation of the expected effects in vivo by comparing the profiles to data from well-characterized psychoactive substances. Comprehensive studies in animal models have been conducted on some of the more popular synthetic cathinones, and they have mostly confirmed that the in vitro data can adequately predict in vivo effects (Aarde and Taffe 2017; Baumann et al. 2017; Green et al. 2014).

6.3.1 Advantages of In Vitro Pharmacological Assessment

There are several clear advantages for initially characterizing new synthetic cathinones using in vitro tests. Foremost, the determination of pharmacological profiles in vitro can be done relatively quickly in laboratories with well-established assays. Large numbers of compounds can therefore be tested, allowing for quick reactions to new substances emerging on the illicit drug market. In heterologous expression systems (see methodological details above), pharmacological potencies can be assessed with high selectivity, since only one target protein at a time is expressed. Importantly, the pharmacological profiles can be determined using human versions of the target proteins. This can be especially relevant when species differences between rodent and human targets have been observed (Barker and Blakely 1996; Han and Gu 2006; Simmler et al. 2016).

6.3.2 Limitations of In Vitro Pharmacological Assessment

Typically, laboratories determine EC_{50} values (often also referred to as IC_{50} if the drug effect is inhibition) for in vitro assessment of data. EC_{50} values depend on assay conditions and are therefore not directly comparable between laboratories. K_i values are comparable, but determining a K_i value requires prior knowledge of the mechanism of action (competitive, noncompetitive, and mixed inhibition) at the target (Burlingham and Widlanski 2003). The inclusion of reference compounds (well-known psychostimulants such as MDMA and methamphetamine) in the screening of NPS can help to compare EC_{50}-based findings between laboratories.

Pharmacological in vitro data can only describe the activity of the mother compound but cannot detect the contributions of potentially bioactive metabolites. For example, MDMA is converted into the active metabolite 3,4-methylenedioxyamphetamine (MDA) (de la Torre et al. 2004; Hysek et al. 2012), which contributes to the overall effects of MDMA administration. As such, by testing compounds in vitro, we cannot exclude the possibility that these compounds are converted into active metabolites that could result in a different relative profile at the targets and affect the predicted psychoactive effects.

6.3.3 Well-Known Psychoactive Substances as Reference Compounds

The most significant element in meaningful in vitro screenings for NPS is the inclusion of well-known psychostimulants as reference compounds. Such substances are typically MDMA, cocaine, or methamphetamine. Synthetic cathinones are derivatives of amphetamine, as are MDMA and methamphetamine. As a tropane alkaloid, cocaine is structurally different than amphetamine derivatives, but with respect to its pharmacological action at monoamine targets, it is a good reference compound. LSD is a hallucinogenic 5-HT_{2A} receptor agonist and can also be used as a reference compound. These reference compounds are ideal not only due to the wealth of knowledge of their molecular mechanisms of action but also due to their diverse pharmacological profiles with respect to their actions at monoamine systems. MDMA is a potent NET and SERT inhibitor with lower potency at DAT.

In addition to inhibiting re-uptake, MDMA also induces transporter-mediated release of NA, 5-HT, and DA (Simmler et al. 2013). Furthermore, it has weak affinity for 5-HT_{2A} receptors (Liechti and Vollenweider 2001; Liechti et al. 2000b; Simmler et al. 2013). Thus, MDMA may serve as the reference compound for drugs that preferentially act on the 5-HT and NA systems. Cocaine nonselectively inhibits the neurotransmitter reuptake at DAT, NET, and SERT (Han and Gu 2006). As a pure uptake inhibitor, cocaine is not a substrate of the transporter and does not induce transporter-mediated release of NA, 5-HT, or DA (Rothman et al. 2001). Methamphetamine is a potent DAT and NET inhibitor. Similar to MDMA, it inhibits neurotransmitter uptake at these transporters by inducing reverse transport through transporter-mediated release (Rothman et al. 2001). Methamphetamine can also inhibit SERT and induce 5-HT release, but only at very high concentrations (Baumann et al. 2012).

6.3.4 Clinical Effects Arising from Elevated Dopaminergic, Serotonergic, and Noradrenergic Transmission

Sympathomimetic stimulation results from activity in the NA system, including NET inhibition, NA release, and stimulation of adrenergic receptors, and is mainly responsible for the psychostimulant and cardiostimulant effects of a drug (Hysek et al. 2011, 2012, 2013). Sympathomimetic stimulant effects are common for cocaine, methamphetamine, and MDMA, as these substances act potently on the noradrenergic system. Cardiostimulant effects induced by LSD are less pronounced than for psychostimulant drugs (Schmid et al. 2015). Agonism of the 5-HT system can result in hyperthermia, hyponatremia, serotonin syndrome, or even seizures (Liechti 2015). 5-HT release also causes entactogenic effects that are typical of MDMA (Liechti et al. 2000a). Hallucinations arise from activation of the 5-HT_{2A} receptor (Kraehenmann et al. 2017; Nichols 2004; Preller et al. 2017). LSD is the prototypical hallucinogenic compound, while hallucinations from MDMA are mild or even absent.

DAT inhibition and DA release both result in elevated DA levels in the mesolimbic system and are characteristic effects of reinforcing substances. Dopaminergic psychostimulants have high abuse and addiction potential due to their reinforcing properties (Luscher and Malenka 2011). Interestingly, the 5-HT system can oppose the reinforcing dopaminergic effects to some extent (Daw et al. 2002). Psychostimulants that are selective for DAT inhibition relative to SERT inhibition showed stronger reinforcing properties in animal models than substances that are selective for SERT inhibition relative to DAT inhibition (Bauer et al. 2013; Wee et al. 2005). To wit, methamphetamine, a highly addictive amphetamine derivative, has high relative potency at DAT over SERT, while MDMA, an amphetamine derivative with low risk for addiction, has high relative potency at SERT over DAT (Simmler et al. 2013). The DAT/SERT ratio (Table 6.1) for

Table 6.1 Uptake inhibition potencies and qualitative releasing properties of selected synthetic cathinones and reference compounds

	Data from	Uptake inhibition potencies			DAT/SERT ratio	Releasing properties		
		NET	DAT	SERT		NA release	DA release	5-HT release
		IC_{50} (μM) (95% CI)	IC_{50} (μM) (95% CI)	IC_{50} (μM) (95% CI)				
Cocaine	(1)	0.451 (0.38–0.59)	0.768 (0.6–1.0)	2.37 (2.0–2.9)	3.1	*N/A*	No	No
MDMA	(1), (3)	0.447 (0.33–0.60)	17 (12–24)	1.36 (1.0–2.0)	0.08	Yes	Yes	Yes
Methamphetamine	(1), (3)	0.064 (0.04–0.09)	1.05 (0.74–1.5)	>10	>10	Yes	Yes	Yes
NET > SERT > DAT selectivity								
2,3-Dimethylmethcathinone	(4)	0.53 (0.36–0.78)	7.4 (5.4–10.1)	1.2 (1.0–1.4)	0.16	Yes	Yes	Yes
2,4-Dimethylmethcathinone	(4)	1.5 (1.1–2.0)	83 (65–105)	1.5 (1.0–2.2)	0.02	Yes	Yes	Yes
3,4-Dimethylmethcathinone	(4)	0.45 (0.33–0.60)	9.4 (7.6–11.7)	1.1 (0.9–1.4)	0.12	Yes	Yes	No
4-Ethylmethcathinone	(3)	2.5 (1.7–3.7)	31 (13–72)	4.3 (3.2–5.9)	0.14	Yes	Yes	Yes
Methedrone	(2)	2.24 (1.4–3.5)	35 (15–79)	4.73 (3.2–6.9)	0.14	Yes	No	Yes
4-Methylmethcathinone	(4)	0.26 (0.19–0.35)	5.7 (4.5–7.2)	3.6 (2.8–4.6)	0.63	Yes	Yes	Yes
4-Bromomethcathinone	(3)	0.41 (0.30–0.57)	5.6 (2.7–12)	2.2 (1.7–2.8)	0.4	Yes	Yes	No
NET > DAT ≥ SERT selectivity								
Buthylone	(1)	2.02 (1.5–2.7)	2.90 (2.5–3.4)	6.22 (4.3–9.0)	2.1	*N/A*	No	Yes
Ethylone	(1)	2.54 (2.0–3.2)	5.68 (4.9–6.5)	4.46 (3.8–5.2)	0.8	*N/A*	No	Yes
Mephedrone	(1), (3)	0.254 (0.22–0.30)	3.31 (2.6–4.2)	4.64 (3.7–5.9)	1.4	Yes	Yes	Yes
4-Methylethcathinone	(2)	2.23 (1.6–3.2)	4.28 (3.4–5.4)	7.93 (3.5–18)	1.85	No	No	Yes
3-Methylmethcathinone	(4)	0.27 (0.21–0.36)	2.6 (2.0–3.3)	9.5 (6.9–13.2)	3.7	Yes	Yes	Yes
Methylone	(1)	0.542 (0.39–0.75)	4.82 (3.8–6.1)	15.5 (10–26)	3.3	*N/A*	No	Yes
Naphyrone	(1), (3)	0.25 (0.20–0.32)	0.47 (0.40–0.55)	0.96 (0.85–1.09)	2.0	No	No	No
Pentylone	(2)	0.99 (0.72–1.4)	1.34 (1.0–1.7)	8.37 (5.4–13)	6.2	No	No	Yes
NET > DAT ≫ SERT selectivity								
Buphedrone	(2)	0.65 (0.51–0.81)	4.24 (3.3–5.5)	70 (2–2700)	>10	Yes	No	No

(continued)

Table 6.1 (continued)

	Data from	Uptake inhibition potencies			DAT/SERT ratio	Releasing properties		
		NET	DAT	SERT		NA release	DA release	5-HT release
		IC_{50} (μM) (95% CI)	IC_{50} (μM) (95% CI)	IC_{50} (μM) (95% CI)				
Cathinone	(1)	0.199 (0.15–0.26)	14.0 (10–20)	>100	>10	*N/A*	Yes	No
N,N-Dimethylcathinone	(2)	7.71 (5–12)	27 (21–36)	>500	>10	No	No	No
Ethcathinone	(2)	0.44 (0.34–0.56)	5.00 (3.7–6.8)	48 (4–529)	9.6	Yes	No	Yes
Flephedrone	(1), (3)	0.246 (0.16–0.37)	6.35 (4.2–9.5)	>10	5.8	Yes	Yes	Yes
3-Fluoromethcathinone	(2)	0.19 (0.13–0.29)	1.7 (1.0–3.0)	56 (7–472)	>10	Yes	Yes	Yes
Methcathinone	(1), (3)	0.085 (0.06–0.17)	1.12 (0.83–1.5)	>10	>10	Yes	Yes	Yes*
Pentedrone	(2)	0.61 (0.52–0.72)	2.50 (2.0–3.2)	135 (5–3700)	>10	No	No	No
Highly potent NET and DAT blockers								
MDPBP	(3)	0.16 (0.11–0.24)	0.11 (0.07–0.16)	15 (5.4–39)	132	No	No	No
MDPPP	(3)	0.97 (0.62–1.5)	0.53 (0.27–1.1)	75 (49–114)	141	No	No	No
MDPV	(1), (3)	0.044 (0.03–0.07)	0.031 (0.03–0.04)	9.30 (6.8–12.8)	>100	No	No	No
Pyrovalerone	(1), (3)	0.043 (0.03–0.06)	0.035 (0.03–0.04)	13.0 (10.8–15.8)	>100	No	No	No
α-PVP	(3)	0.02 (0.01–0.03)	0.04 (0.01–0.1)	>100	>1000	No	No	No

(1) Simmler et al. (2013); (2) Simmler et al. (2014); (3) Rickli et al. (2015); (4) Lüthi et al. (2017); (*N/A*) not assessed; (*) Not significant in Rickli et al. (2015)

relative uptake inhibition potency, calculated as 1/DAT IC_{50}:1/SERT IC_{50}, can serve to predict the abuse liability of novel synthetic cathinones. In acute intoxications, dopaminergic action can result in confusion and psychotic states, which is often observed during intoxication with synthetic cathinones (White 2016).

6.4 Pharmacological Diversity of Synthetic Cathinones

All synthetic cathinones are structurally related through the characteristic β-keto group at the carbon side chain, but their pharmacological profiles and potencies can vary considerably (Fig. 6.3). Such variability is also observed among non-β-keto amphetamine derivatives, such as MDMA versus methamphetamine. In contrast, potent uptake inhibition at NET is a shared characteristic of almost all of the different psychostimulants, including synthetic cathinones (with some exceptions that only moderately inhibit NET, for example *N,N*-dimethylcathinone, ethylone, methedrone, and 4-methylethcathinone; see Table 6.1). Therefore, it seems that the differences in the pharmacological profiles arise mostly from drug actions on the DA- and 5-HT systems. We use the relative potencies at DAT and SERT (DAT/SERT ratio) to determine the similarity to the reference compounds MDMA, methamphetamine, and cocaine. Tables 6.1 and 6.2 provide an overview of the pharmacological profiles of selected synthetic cathinones. Clinically relevant differences between different synthetic cathinone pharmacological profiles are discussed in Sects. 6.4.1–6.4.3.

This discussion will cover three main groups: first, synthetic cathinones that act with high potency at SERT and cause release of 5-HT can be expected to have subjective effects and intoxication profiles that are similar to MDMA or cocaine. Second, synthetic cathinones acting preferentially at the noradrenergic and dopaminergic systems can be compared to methamphetamine and have high potential for abuse due to the expected reinforcing effects. Third, special caution regarding risk for severe acute overdosing is needed for the extremely potent DAT and NET blockers, which are significantly more potent uptake inhibitors than cocaine and should be discussed separately.

6.4.1 Synthetic Cathinones with Selective and Nonselective Serotonergic Action

There are a few synthetic cathinones that resemble MDMA in their selectivity for SERT. For example, methedrone (Simmler et al. 2014), 4-ethylmethcathinone (Rickli et al. 2015), 2,3-dimethylmethcathinone, 3,4-dimethylmethcathinone, and 2,4-dimethylmethcathinone (Luethi et al. 2017) all inhibit SERT with high selectivity over DAT

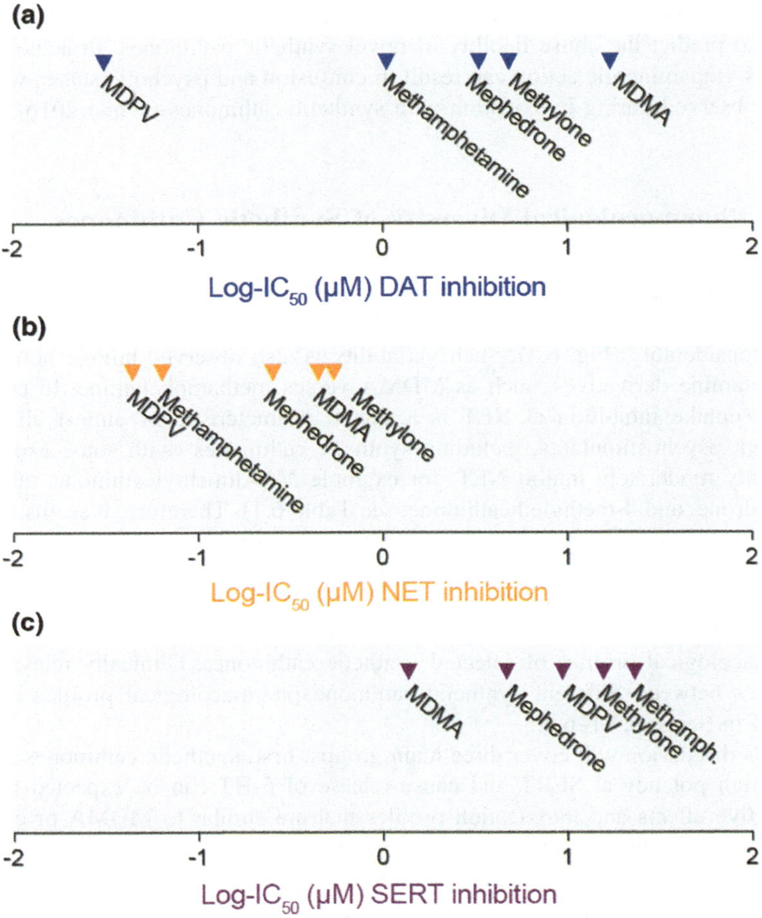

Fig. 6.3 Variation in pharmacological uptake inhibition profiles between synthetic cathinones and the reference compounds MDMA and methamphetamine. Log-transformed IC_{50} values for uptake inhibition at DAT (**a**), NET (**b**), and SERT (**c**) are illustrated for the synthetic cathinones mephedrone, methylone, and MDPV. The amphetamine derivatives MDMA and methamphetamine are included for comparison. IC_{50} values are from Simmler et al. (2013)

inhibition, comparable to MDMA. 4-Bromomethcathinone (Eshleman et al. 2017; Rickli et al. 2015) and 4-methylmethcathinone (Luethi et al. 2017) are also potent SERT inhibitors, comparable to MDMA; however, unlike MDMA and the previously mentioned synthetic cathinones, they are also fairly potent DAT inhibitors. With the exception of 3,4-dimethylmethcathinone, all of these serotonergic cathinones also act as 5-HT releasers. 4-Ethylmethcathinone, 4-bromomethcathinone (Rickli et al. 2015),

4-methylmethcathinone, 2,3-dimethylmethcathinone, 3,4-dimethylmethcathinone, and 2,4-dimethylmethcathinone (Luethi et al. 2017) showed binding to 5-HT$_{2A}$ receptors with similar or higher affinity than MDMA (Rickli et al. 2015; Table 6.2). In activity assays, 2,3-dimethylmethcathinone and 4-methylmethcatinone stood out as good 5-HT$_{2A}$ receptor activators, suggesting they may cause hallucinations. The serotonergic properties of these particular cathinones are rather rare among the synthetic cathinones in general. The β-keto group that is characteristic of cathinones seems to generally reduce SERT activity, as evidenced by comparison to amphetamine derivatives that are structurally identical to synthetic cathinones except that they lack the β-keto group. For example, methylone, which is the "cathinone version" of MDMA, is over tenfold less potent in inhibiting SERT than MDMA but has a similar activity at NET and around a threefold higher potency at DAT (Eshleman et al. 2013; Simmler et al. 2013).

Many cathinone derivatives show nonselective inhibition of SERT, DAT, and NET, similar to cocaine. However, unlike cocaine which acts as pure uptake inhibitor, most of these nonspecific cathinones also induce transporter-mediated release. Nevertheless, in terms of acute psychotropic effects, acute intoxication, and abuse liability, these cathinones can be expected to have similar effects to cocaine. Mephedrone, methylone, ethylone, butylone, and naphyrone are nonselective synthetic cathinones (Simmler et al. 2016). 4-Methylethcathinone and pentylone also act nonselectively at SERT and DAT (Saha et al. 2015; Simmler et al. 2014). Interestingly, 4-methylethcathinone is a "releaser/blocker hybrid" since it induces 5-HT, but not DA release.

In general, synthetic cathinones have poor affinity at the trace amine-associated receptor 1 (TAAR1). Exceptions are 2,4-dimethylmethcathinone and 2,3-dimethylmethcathinone, which have affinities in the sub-micromolar range for rat and mouse TAAR1, respectively (Luethi et al. 2017). TAAR1 is involved in the regulation of DA neuron activity, and its activation can reduce the psychostimulant effects of a drug (Bradaia et al. 2009; Revel et al. 2012). However, due to species differences that are typically observed for psychostimulant actions at TAAR1 (Simmler et al. 2016), it is unclear if the activity at TAAR1 induced by 2,4-dimethylmethcathinone and 2,3-dimethylmethcathinone would play a significant role in humans.

6.4.2 Dopaminergic and Noradrenergic Cathinones

Similar to the synthetic cathinones described above, the compounds summarized here are also potent NET inhibitors, and many of them also act as NA releasers. Several synthetic cathinones also show weak affinity at α_{1A}- and α_{2A}-adrenergic receptors (Table 6.2).

Cathinone, methcathinone, flephedrone (Simmler et al. 2013), ethcathinone, 3-fluoromethcathinone, and buphedrone (Simmler et al. 2014) are typical synthetic cathinones with selective action at DAT compared to SERT (Table 6.1). Typically, these DAT- and NET-selective cathinones act as DA and NA releasers. Buphedrone, however, releases NA but not DA (Simmler et al. 2014), and

Table 6.2 Binding affinities at selected monoamine receptors

	Data from	Receptor binding affinity K_i (μM), mean ± SD				
		5-HT_{1A}	5-HT_{2A}	5-HT_{2C}	α_{1A}	α_{2A}
Reference compounds						
Cocaine	(1)	>20	>13	>13	>6	>20
MDMA	(1)	12.2 ± 0.8	7.8 ± 2.4	>13	>6	15.0 ± 10
Methamphetamine	(1)	8.07 ± 0.75	>13	>13	>6	6.1 ± 1.6
NET > SERT > DAT selectivity for uptake *inhibition*						
2,3-Dimethylmethcathinone	(4)	>17	0.64 ± 0.19	2.4 ± 0.9	0.78 ± 0.10	3.0 ± 0.1
2,4-Dimethylmethcathinone	(4)	15 ± 3	1.3 ± 0.1	1.3 ± 0.3	0.16 ± 0.02	3.0 ± 0.3
3,4-Dimethylmethcathinone	(4)	>17	1.9 ± 0.3	1.5 ± 0.2	1.9 ± 0.3	3.5 ± 0.2
4-Ethylmethcathinone	(3)	>20	6.5 ± 0.9	9.6 ± 0.4	8.4 ± 3.4	21.1 ± 7.6
Methedrone	(2)	>20	>13	>13	>6	>25
4-Methylmethcathinone	(4)	>17	1.6 ± 0.2	8.1 ± 5.4	1.1 ± 0.1	11 ± 1
4-Bromomethcathinone	(3)	>20	3.2 ± 0.6	>13	8.2 ± 3.0	12.7 ± 0.2
NET > DAT ≥ SERT selectivity for uptake inhibition						
Buthylone	(1)	>20	>13	>13	>6	>25
Ethylone	(1)	17.0 ± 2.4	>13	>13	>6	>25
Mephedrone	(1)	>20	2.1 ± 0.7	>13	3.48 ± 2.2	11.0 ± 5.0
4-Methylethcathinone	(2)	>20	3.8 ± 1.6	5.2 ± 0.3	>6	>18
3-Methylmethcathinone	(4)	4.8 ± 0.5	3.4 ± 0.8	3.6 ± 1.0	7.9 ± 0.2	1.1 ± 0.1
Methylone	(1)	>20	>13	>13	>6	>20
Naphyrone	(1)	6.00 ± 0.21	11 ± 2.2	>13	>6	8.0 ± 2.8
Pentylone	(2)	>18	>13	>13	>6	>25
NET > DAT ≫ SERT selectivity for uptake inhibition						
Buphedrone	(2)	>18	>13	>13	>6	23.9 ± 4.2

(continued)

Table 6.2 (continued)

	Data from	Receptor binding affinity K_i (μM), mean ± SD				
		5-HT_{1A}	5-HT_{2A}	5-HT_{2C}	α_{1A}	α_{2A}
Cathinone	(1)	>20	>13	>13	5.40 ± 1.1	8.9 ± 2.7
N,N-Dimethylcathinone	(2)	>18	6.5 ± 0.8	6.5 ± 0.8	>6	25.4 ± 11
Ethcathinone	(2)	8.5 ± 1.1	>13	9.3 ± 0.2	>6	15.5 ± 1.9
Flephedrone	(1)	>20	1.4 ± 0.6	>13	1.52 ± 0.05	>20
3-Fluoromethcathinone	(2)	>18	>13	6.1 ± 2.2	>6	10 ± 2.2
Methcathinone	(1)	12.7 ± 3.5	3.0 ± 0.6	>13	3.93 ± 1.3	11.9 ± 3.9
Pentedrone	(2)	>18	>13	>13	>6	35.4 ± 16
Highly potent NET and DAT blockers						
MDPBP	(3)	13.0 0.02	>13	>13	>4.9	9.4 ± 1.6
MDPPP	(3)	2.5 ± 0.3	7.5 ± 0.1	>13	>15	13.9 ± 0.9
MDPV	(1)	10.29 ± 4.7	>13	>13	>6	>20
Pyrovalerone	(1)	13.4 ± 2.1	>13	>13	>6	>20
α-PVP	(3)	5.2 ± 0.1	>13	>13	>15	>20

(1) Simmler et al. (2013); (2) Simmler et al. (2014); (3) Rickli et al. (2015); (4) Lüthi et al. (2017)

pentedrone and *N,N*-dimethylcathinone only act as pure DAT and NET blockers and not releasers (Simmler et al. 2014). The selectivity for dopaminergic versus serotonergic action indicates a high risk for abuse of these substances, comparable to methamphetamine (Wee et al. 2005). Interestingly, flephedrone showed affinity for 5-HT_{2C} receptor, perhaps mitigating its abuse liability (Table 6.2).

6.4.3 Potent Uptake Inhibitors

Synthetic cathinones with a pyrovalerone chemical structure are typically very potent blockers of DAT and NET, but they lack the neurotransmitter releasing properties (Table 6.1). They also have low or negligible potency for SERT inhibition (Meltzer et al. 2006; Zawilska and Wojcieszak 2017). The length of the carbon chain at the α-position seems to be favorable for potent DAT inhibition (Eshleman et al. 2017), while a bulky substitution of the amine seems to prevent release characteristics (Shalabi et al. 2017). 3,4-Methylenedioxypyrovalerone (MDPV), α–pyrrolidinovalerophenone (α–PVP), 3,4-methylenedioxy-α-pyrrolidinopropiophenone (MDPPP), and 3,4-methylenedioxy-α-pyrrolidinobutiophenone (MDPBP) are typical pyrovalerone cathinones with weak or negligible SERT inhibition, but DAT and NET inhibition potencies in the nanomolar range (Rickli et al. 2015; Simmler et al. 2014). There are a few exceptions among pyrovalerone cathinones with respect to SERT activity. For example, naphyrone (Simmler et al. 2013) and the 4-I-derivative of pyrovalerone (Meltzer et al. 2006) inhibit SERT relatively potently. MDPV and α-PVP are extremely potent DAT and NET blockers, comparable to pyrovalerone (Rickli et al. 2015). Pyrovalerone cathinones do not show significant affinity for monoamine receptors (Table 6.2). Due to the high potencies of pyrovalerone cathinones, overdosing is very likely. Severe and long-lasting intoxication has been described, as have fatalities (Zawilska and Wojcieszak 2017). In vitro assessments of the pharmacology of these drugs are of particular value here, since extremely potent, and therefore particularly dangerous, psychostimulants can be identified quickly.

6.5 Conclusion

Synthetic cathinones vary considerably in their pharmacological profiles. In vitro, most of the synthetic cathinones act as NET inhibitors, but the selectivity for DAT *versus* SERT inhibition varies among substances. Synthetic cathinones with less bulky chemical structures are not only transporter blockers but also induce neurotransmitter release via reverse transport. In contrast, pyrovalerone cathinones are pure and very potent DAT and NET blockers. The toxicity and acute effects of synthetic cathinones can be estimated from in vitro profiling and comparison to well-known reference compounds such as MDMA, methamphetamine, and cocaine. Synthetic cathinones tend toward selective inhibition of DAT over SERT, which indicates high abuse liability. Pyrovalerone cathinones are significantly more potent drugs than conventional psychostimulants, bearing substantial risk for overdoses.

Acknowledgements The author would like to thank Dr. Shannon L. Wolfman for proofreading the manuscript. The author is supported by the Swiss National Science Foundation (PZ00P3_174178).

References

Aarde SM, Taffe MA (2017) Predicting the abuse liability of entactogen-class, new and emerging psychoactive substances via preclinical models of drug self-administration. Curr Top Behav Neurosci 32:145–164. https://doi.org/10.1007/7854_2016_54

Barker EL, Blakely RD (1996) Identification of a single amino acid, phenylalanine 586, that is responsible for high affinity interactions of tricyclic antidepressants with the human serotonin transporter. Mol Pharmacol 50(4):957–965

Bauer CT, Banks ML, Blough BE, Negus SS (2013) Use of intracranial self-stimulation to evaluate abuse-related and abuse-limiting effects of monoamine releasers in rats. Br J Pharmacol 168(4):850–862. https://doi.org/10.1111/j.1476-5381.2012.02214.x

Baumann MH, Ayestas MA Jr, Partilla JS et al (2012) The designer methcathinone analogs, mephedrone and methylone, are substrates for monoamine transporters in brain tissue. Neuropsychopharmacology 37(5):1192–1203. https://doi.org/10.1038/npp.2011.304

Baumann MH, Bukhari MO, Lehner KR et al (2017) Neuropharmacology of 3,4-methylenedioxypyrovalerone (MDPV), its metabolites, and related analogs. Curr Top Behav Neurosci 32:93–117. https://doi.org/10.1007/7854_2016_53

Blakely RD, Defelice LJ, Galli A (2005) Biogenic amine neurotransmitter transporters: just when you thought you knew them. Physiology 20:225–231. https://doi.org/10.1152/physiol.00013.2005

Bradaia A, Trube G, Stalder H et al (2009) The selective antagonist EPPTB reveals TAAR1-mediated regulatory mechanisms in dopaminergic neurons of the mesolimbic system. Proc Natl Acad Sci U S A 106(47):20081–20086. https://doi.org/10.1073/pnas.0906522106

Burlingham BT, Widlanski TS (2003) An intuitive look at the relationship of K-i and IC50: a more general use for the Dixon plot. J Chem Educ 80(2):214–218

Daw ND, Kakade S, Dayan P (2002) Opponent interactions between serotonin and dopamine. Neural Net 15(4–6):603–616

de la Torre R, Farre M, Roset PN et al (2004) Human pharmacology of MDMA: pharmacokinetics, metabolism, and disposition. Ther Drug Monit 26(2):137–144

EMCDDA (2016) European drug report 2016. European Monitoring Centre for Drugs and Drug Addiction (EMCDDA), Luxembourg. https://doi.org/10.2810/084165

Eshleman AJ, Wolfrum KM, Hatfield MG et al (2013) Substituted methcathinones differ in transporter and receptor interactions. Biochem Pharmac 85(12):1803–1815. https://doi.org/10.1016/j.bcp.2013.04.004

Eshleman AJ, Wolfrum KM, Reed JF et al (2017) Structure-activity relationships of substituted cathinones, with transporter binding, uptake, and release. J Pharmacol Exp Ther 360(1):33–47. https://doi.org/10.1124/jpet.116.236349

Green AR, King MV, Shortall SE, Fone KC (2014) The preclinical pharmacology of mephedrone; not just MDMA by another name. Br J Pharmacol 171(9):2251–2268. https://doi.org/10.1111/bph.12628

Gregg RA, Baumann MH, Partilla JS et al (2015) Stereochemistry of mephedrone neuropharmacology: enantiomer-specific behavioural and neurochemical effects in rats. Br J Pharmacol 172(3):883–894. https://doi.org/10.1111/bph.12951
Han DD, Gu HH (2006) Comparison of the monoamine transporters from human and mouse in their sensitivities to psychostimulant drugs. BMC Pharmacol 6:6. https://doi.org/10.1186/1471-2210-6-6
Hysek CM, Simmler LD, Ineichen M et al (2011) The norepinephrine transporter inhibitor reboxetine reduces stimulant effects of MDMA ("ecstasy") in humans. Clin Pharmacol Ther 90 (2):246–255. https://doi.org/10.1038/clpt.2011.78
Hysek CM, Simmler LD, Nicola VG et al (2012) Duloxetine inhibits effects of MDMA ("Ecstasy") in vitro and in humans in a randomized placebo-controlled laboratory study. PLoS ONE 7(5):e36476. https://doi.org/10.1371/journal.pone.0036476
Hysek CM, Fink AE, Simmler LD et al (2013) α_1-Adrenergic receptors contribute to the acute effects of 3,4-methylenedioxymethamphetamine in humans. J Clin Psychopharmacol 33 (5):658–666. https://doi.org/10.1097/JCP.0b013e3182979d32
Iversen L, Gibbons S, Treble R et al (2013) Neurochemical profiles of some novel psychoactive substances. Eur J Pharmacol 700(1–3):147–151. https://doi.org/10.1016/j.ejphar.2012.12.006
Kraehenmann R, Pokorny D, Vollenweider L et al (2017) Dreamlike effects of LSD on waking imagery in humans depend on serotonin 2A receptor activation. Psychopharmacology 234 (13):2031–2046. https://doi.org/10.1007/s00213-017-4610-0
Liechti M (2015) Novel psychoactive substances (designer drugs): overview and pharmacology of modulators of monoamine signaling. Swiss Med Wkly 145:w14043. https://doi.org/10.4414/smw.2015.14043
Liechti ME, Vollenweider FX (2001) Which neuroreceptors mediate the subjective effects of MDMA in humans? A summary of mechanistic studies. Hum Psychopharmacol 16(8):589–598. https://doi.org/10.1002/hup.348
Liechti ME, Baumann C, Gamma A, Vollenweider FX (2000a) Acute psychological effects of 3,4-methylenedioxymethamphetamine (MDMA, "Ecstasy") are attenuated by the serotonin uptake inhibitor citalopram. Neuropsychopharmacology 22(5):513–521. https://doi.org/10.1016/S0893-133X(99)00148-7
Liechti ME, Saur MR, Gamma A et al (2000b) Psychological and physiological effects of MDMA ("Ecstasy") after pretreatment with the 5-HT_2 antagonist ketanserin in healthy humans. Neuropsychopharmacology 23(4):396–404. https://doi.org/10.1016/S0893-133X(00)00126-3
Luethi D, Kolaczynska KE, Docci L et al (2017) Pharmacological profile of mephedrone analogs and related new psychoactive substances. Neuropharmacology. https://doi.org/10.1016/j.neuropharm.2017.07.026
Luscher C, Malenka RC (2011) Drug-evoked synaptic plasticity in addiction: from molecular changes to circuit remodeling. Neuron 69(4):650–663. https://doi.org/10.1016/j.neuron.2011.01.017
Meltzer PC, Butler D, Deschamps JR, Madras BK (2006) 1-(4-Methylphenyl)-2-pyrrolidin-1-yl-pentan-1-one (pyrovalerone) analogues: a promising class of monoamine uptake inhibitors. J Med Chem 49(4):1420–1432. https://doi.org/10.1021/jm050797a
Nichols DE (2004) Hallucinogens. Pharmacol Ther 101(2):131–181. https://doi.org/10.1016/j.pharmthera.2003.11.002
Nickell JR, Siripurapu KB, Vartak A et al (2014) The vesicular monoamine transporter-2: an important pharmacological target for the discovery of novel therapeutics to treat methamphetamine abuse. Adv Pharmacol 69:71–106. https://doi.org/10.1016/B978-0-12-420118-7.00002-0
Preller KH, Herdener M, Pokorny T et al (2017) The fabric of meaning and subjective effects in LSD-induced states depend on serotonin 2A receptor activation. Current Biol 27(3):451–457. https://doi.org/10.1016/j.cub.2016.12.030
Revel FG, Meyer CA, Bradaia A et al (2012) Brain-specific overexpression of trace amine-associated receptor 1 alters monoaminergic neurotransmission and decreases sensitivity

to amphetamine. Neuropsychopharmacology 37(12):2580–2592. https://doi.org/10.1038/npp.2012.109

Rickli A, Hoener MC, Liechti ME (2015) Monoamine transporter and receptor interaction profiles of novel psychoactive substances: para-halogenated amphetamines and pyrovalerone cathinones. Eur Neuropsychopharmacol 25(3):365–376. https://doi.org/10.1016/j.euroneuro.2014.12.012

Rothman RB, Baumann MH, Dersch CM et al (2001) Amphetamine-type central nervous system stimulants release norepinephrine more potently than they release dopamine and serotonin. Synapse 39(1):32–41. https://doi.org/10.1002/1098-2396(20010101)39:1<32:AID-SYN5>3.0.CO;2-3

Saha K, Partilla JS, Lehner KR et al (2015) 'Second-generation' mephedrone analogs, 4-MEC and 4-MePPP, differentially affect monoamine transporter function. Neuropsychopharmacology 40(6):1321–1331. https://doi.org/10.1038/npp.2014.325

Schmid Y, Enzler F, Gasser P et al (2015) Acute effects of lysergic acid diethylamide in healthy subjects. Biol Psychiatry 78(8):544–553. https://doi.org/10.1016/j.biopsych.2014.11.015

Scholze P, Zwach J, Kattinger A et al (2000) Transporter-mediated release: a superfusion study on human embryonic kidney cells stably expressing the human serotonin transporter. J Pharmacol Exp Ther 293(3):870–878

Shalabi AR, Walther D, Baumann MH, Glennon RA (2017) Deconstructed analogues of bupropion reveal structural requirements for transporter inhibition versus substrate-induced neurotransmitter release. ACS Chem Neurosci 8(6):1397–1403. https://doi.org/10.1021/acschemneuro.7b00055

Simmler LD, Liechti ME (2017) Interactions of cathinone NPS with human transporters and receptors in transfected cells. Curr Top Behav Neurosci 32:49–72. https://doi.org/10.1007/7854_2016_20

Simmler L, Buser T, Donzelli M et al (2013) Pharmacological characterization of designer cathinones in vitro. Br J Pharmacol 168(2):458–470. https://doi.org/10.1111/j.1476-5381.2012.02145.x

Simmler LD, Rickli A, Hoener MC, Liechti ME (2014) Monoamine transporter and receptor interaction profiles of a new series of designer cathinones. Neuropharmacology 79:152–160. https://doi.org/10.1016/j.neuropharm.2013.11.008

Simmler LD, Buchy D, Chaboz S et al (2016) In vitro characterization of psychoactive substances at rat, mouse, and human trace amine-associated receptor 1. J Pharmacol Exp Ther 357(1):134–144. https://doi.org/10.1124/jpet.115.229765

Wee S, Anderson KG, Baumann MH et al (2005) Relationship between the serotonergic activity and reinforcing effects of a series of amphetamine analogs. J Pharmacol Exp Ther 313(2):848–854. https://doi.org/10.1124/jpet.104.080101

White CM (2016) Mephedrone and 3,4-methylenedioxypyrovalerone (MDPV): synthetic cathinones with serious health implications. J Clin Pharmacol 56(11):1319–1325. https://doi.org/10.1002/jcph.742

Zawilska JB, Wojcieszak J (2017) α-Pyrrolidinophenones: a new wave of designer cathinones. Forensic Toxicol 35(2):201–216. https://doi.org/10.1007/s11419-016-0353-6

Chapter 7
Effects of Synthetic Cathinones on Brain Neurotransmitters

Krystyna Gołembiowska and Katarzyna Kamińska

Abstract Synthetic cathinones are the β-keto analogs of amphetamines and synthetic derivatives of cathinone naturally present in the khat plant (*Catha edulis*). There is very little data on pharmacology of cathinone derivatives. Like amphetamines, cathinones exert their stimulant effects via increasing synaptic concentrations of dopamine (DA), noradrenaline (NA), and serotonin (5-HT), albeit at various potencies. These compounds interact with plasma membrane transporters for DA, NA, and 5-HT. Some designer cathinones target monoamine transporters like substrates, such as amphetamine, while others are their blockers, like cocaine. Substrates, but not blockers, stimulate non-exocytotic release of neurotransmitters by reversing the normal direction of the transporter flux. Blockers of transport proteins are capable of increasing the extracellular level of neurotransmitters by preventing their reuptake after release from monoaminergic neurons. This chapter provides an overview of effects of synthetic cathinones on brain neurotransmitter levels. In vivo microdialysis studies demonstrated that administration of cathinones produced a potent and dose-dependent increase in the extracellular DA and 5-HT levels in the rat nucleus accumbens, frontal cortex, and striatum. It appears that certain synthetic cathinones potently affect extracellular levels of brain monoaminergic neurotransmitters, which may contribute to psychostimulatory, entactogenic, and hallucinogenic effect of these drugs and brings about the risk of addiction.

Keywords Dopamine · Serotonin · Release · Microdialysis

K. Gołembiowska (✉) · K. Kamińska
Institute of Pharmacology, Polish Academy of Sciences, Smętna 12,
31-343 Kraków, Poland
e-mail: nfgolemb@cyf-kr.edu.pl

J. B. Zawilska (ed.), *Synthetic Cathinones*, Current Topics in Neurotoxicity 12,
https://doi.org/10.1007/978-3-319-78707-7_7

7.1 Introduction

Synthetic cathinone derivatives exert their stimulant effects through the interaction with monoamine membrane transporters, namely for dopamine (DAT), serotonin (SERT), and noradrenaline (NET). Cathinone derivatives are closely related to phenylethylamine stimulants, like MDMA and methamphetamine, and differ by the presence of a β-keto group in the aliphatic side chain. Due to this structural analogy, synthetic cathinones exhibit central nervous system stimulant effects similar to those observed after amphetamines.

β-Ketoamphetamines have been classified as substrates and non-substrates based on whether or not they are transported by DAT. For instance, mephedrone and methylone, like methamphetamine, are substrates for DAT-mediated uptake and they cause DA release via carrier-mediated exchange (Baumann et al. 2012; Cameron et al. 2013; Eshleman et al. 2013). In contrast, 3,4-methylenedioxypyrovalerone (MDPV) is not a substrate for the transporter protein and interacts with DAT strictly as a blocker, like cocaine (Baumann et al. 2012; Cameron et al. 2013; Eshleman et al. 2013; Simmler et al. 2013). These properties represent an important mechanism by which cathinones exert their effects on release of neurotransmitters from nerve endings. Using microdialysis in conscious rats, a number of studies showed that depending on the substituent in *para* position of the phenyl unit of methcathinone, the compound produces different effects on extracellular DA or 5-HT level.

7.2 Effects of Methcathinone, Methylone, and Mephedrone on Extracellular DA and 5-HT Levels in Rat Brain Regions

In the early work of Gygi et al. (1997), methcathinone (1, 10, 20, and 30 mg/kg, s.c.) dose-dependently increased extracellular striatal DA level in rats. Interestingly, the maximal increases in striatal DA at the three higher doses did not differ significantly. In a later study by Suyama et al. (2016), methcathinone (0.32, 1 and 3.2 mg/kg, i.p.) significantly increased extracellular 5-HT but not DA level in the rat nucleus accumbens.

Methylone and its 4-methyl-substituted analog, mephedrone (0.3 and 1 mg/kg, i.v.) produced concurrent elevations in extracellular DA and 5-HT in vivo in the rat nucleus accumbens. The relative impact of both drugs was greater on dialysate 5-HT (Baumann et al. 2012). Methylone and mephedrone produced elevations in dialysate concentrations of DA and 5-HT qualitatively analogous to the effect of MDMA. This is in contrast to the effect of methamphetamine, which caused preferential increase in DA rather that 5-HT (Baumann et al. 2012). Since extracellular level of DA in the nucleus accumbens is critically linked to the reinforcing effect of abused drugs, it may be predicted that methylone and mephedrone would be reinforcing (Wise 2008).

Further, mephedrone (1 and 3 mg/kg, s.c.) significantly and dose-dependently increased both DA and 5-HT levels in the rat nucleus accumbens (Kehr et al. 2011). The effect of the higher mephedrone dose on 5-HT release was comparable to the effect induced by the same dose of MDMA. However, the percentage of 5-HT increase was at least two times higher for MDMA than for mephedrone. In addition, mephedrone potently increased the accumbal DA release to a level that was comparable to the effect induced by amphetamine, which had only a minor effect on extracellular 5-HT concentrations. These data demonstrate that mephedrone, like MDMA, markedly increases 5-HT release, but is less efficient in enhancing DA release (Kehr et al. 2011).

Another study extended the findings of Kehr et al. (2011) and Baumann et al. (2012), and showed that mephedrone given in single doses of 1, 3.2, and 10 mg/kg (i.p.) significantly altered in vivo DA versus 5-HT selectivity, in a such way that mephedrone preferentially increased 5-HT versus DA levels in the rat nucleus accumbens (Suyama et al. 2016).

Shortall et al. (2015) studied the effect of mephedrone given in a way typical for its recreational use. Three intraperitoneal injections of mephedrone (10 mg/kg) at 2 h intervals increased striatal extracellular DA levels at 40 min after the first and third injections. DA levels returned to basal values in inter-injection intervals. 5-HT level was less affected.

In a more extensive study investigating mephedrone effects on extracellular neurotransmitter levels, it was shown that single doses (5, 10, and 20 mg/kg, i.p.) of mephedrone dose-dependently increased extracellular DA and 5-HT levels in the rat striatum, nucleus accumbens, and frontal cortex (Gołembiowska et al. 2016). In all studied brain regions, the enhancement of 5-HT level was stronger than that of DA. Notably, the effect on 5-HT was the most potent in the striatum and frontal cortex. In comparison to MDMA (5 and 10 mg/kg, i.p.), mephedrone was more efficient in increasing DA level in the striatum and frontal cortex (Gołembiowska et al. 2016). On the other hand, MDMA was more effective in increasing 5-HT in the nucleus accumbens, but was weaker in the striatum. These data are indicative of regional differences in the effect of both drugs and show that mephedrone is a more preferential releaser of 5-HT than DA.

7.3 Effects of 3,4-Methylenedioxypyrovalerone (MDPV) on Extracellular DA and 5-HT Levels in Rat Brain Regions

MDPV as a very potent inhibitor of DAT and NET, at least tenfold more potent than cocaine, does not release monoamines but increases extracellular level of DA in the rat nucleus accumbens when given at doses of 0.1 and 0.3 mg/kg (i.v.)

(Baumann et al. 2013). The potent blockade of DA uptake caused by MDPV is predictive of a high risk for abuse of this drug. However, powerful cocaine-like action of MDPV may be used to manage excessive dopaminergic stimulation.

7.4 Effects of *Para*-Substituted Methcathinone Analogs on Extracellular DA and 5-HT Levels in Rat Brain Regions

A *para* (4-position) substituent on the methcathinone structure is a significant determinant of the neurochemical effects of synthetic cathinones and a predictive factor of in vivo drug effect on monoamines release. Microdialysis study performed in the rat nucleus accumbens with *para*-halogen substitution (-F, -Cl or -Br) showed that halogenated analogs of methcathinone increased DA and 5-HT release with different potencies (Suyama et al. 2016). 4-Fluoromethcathinone (flephedrone) and 4-bromomethcathinone (brephedrone) preferentially increased extracellular 5-HT level, while 4-chloromethcathinone (clephedrone) more efficiently increased extracellular DA level than 5-HT level. Thus, the net effect of 4-F and 4-Br analogs was a decrease in DA versus 5-HT selectivity. 4-Methoxymethcathinone (methedrone) displayed an increased potency to raise 5-HT over DA extracellular level (Suyama et al. 2016).

7.5 Effects of Trifluoromethyl Ring-Substituted Methcathinone Analogs on Extracellular DA and 5-HT Levels in Rat Brain Regions

Meta- and *para*-trifluoromethyl-substituted analogs of methcathinone were weaker NET and DAT inhibitors than methcathinone (Cozzi et al. 2013). However, they produced a dose-dependent (1 and 3 mg/kg, i.v.) increase in 5-HT but not DA release in the rat nucleus accumbens (Cozzi et al. 2013) in contrast to the effect of methcathinone (0.3 and 1.0 mg/kg, i.v.) which stimulated both 5-HT and DA release in the nucleus accumbens. Authors of that work concluded that the response to methcathinone in the in vivo microdialysis study paralleled the in vitro results as methcathinone was a releaser of both 5-HT and DA. On the other hand, *meta-* and *para*-phenyl ring substitution dramatically increased potency at SERT compared to DAT.

7.6 Concluding Remarks

The pharmacological properties of cathinone derivatives depend on their effects on monoamine transporters. The distinction in the molecular mechanism of action of cathinones is essential to determine their impact on synaptic neurotransmitter level. Mephedrone, methylone, and methcathinone as substrates for DAT and SERT are translocated into neurons and evoke non-exocytotic release of transmitters via reverse transport, like MDMA and methamphetamine. Methcathinone exhibits a DAT/SERT inhibition ratio similar to that of methamphetamine, with high inhibitory potencies at DAT and low potencies at SERT (Simmler et al. 2013). Methylone was slightly more potent DAT than SERT inhibitor (Eshleman et al. 2013), while mephedrone was equally potent at DAT and SERT (Hadlock et al. 2011). However, there are some data showing that mephedrone was more potent in increasing extracellular 5-HT level than DA level. These results may be explained by the lack of selectivity of SERT, which is also capable of reuptake of DA into serotonergic nerve terminals. In this way, high intracellular DA concentrations in the serotonin neurons are responsible for indirect overflow of 5-HT into the synaptic cleft (Larsen et al. 2011). On the other hand, transporter blockers by a simple blockade of monoamine carriers increase extracellular monoamine levels similarly to cocaine. MDPV, the selective DAT blocker, increases extracellular DA level, is an example of such substances. Another possible molecular mechanism of cathinone effect on extracellular neurotransmitter level is related to their action through direct activation of serotonin receptors or indirect receptor activation by increased endogenous DA or 5-HT levels. It was shown that cathinone analogs displayed affinity at 5-HT_{1A}, 5-HT_{2A}, 5-HT_{2C}, and D1/D2 receptors (Eshleman et al. 2013; Rickli et al. 2015; Simmler and Liechti 2017). Therefore, it may be suggested that some of the in vivo effects on DA and 5-HT release may be mediated via serotonin or dopamine receptors. Stimulation of 5-HT_{2A} receptors in the ventral tegmental area (VTA) by cathinone analogs may stimulate DA release in the nucleus accumbens or the frontal cortex (Gołembiowska et al. 2016). Furthermore, activation of 5-HT_{2A} receptors located on glutamatergic neurons projecting to nigral or VTA regions may also affect DA or 5-HT release from nerve terminals (Gołembiowska et al. 2016). Alternatively, postsynaptic effect via D1/D2 dopamine receptors situated in medium spiny GABAergic neurons may also indirectly influence monoamine synthesis and release in mesolimbic and mesocortical brain regions.

Table 7.1 The summary of the in vivo cathinone analogs effect on neurotransmitter release

	DA	5-HT	References
Methcathinone			
1, 10, 20, 30 mg/kg s.c. 0.3, 1, 3.2 mg/kg i.p. 0.3, 1 mg/kg i.v.	↑(striatum) ↑↑(n. accumbens) ↑↑ (n. accumbens)	Not tested ↑ ↑	Gygi et al. (1997) Suyama et al. (2016) Cozzi et al. (2013)
Mephedrone			
0.3, 1 mg/kg i.v. 1, 3 mg/kg s.c. 1, 3.2, 10 mg/kg i.p. 10 mg/kg × 3 i.p. 5, 10, 20 mg/kg i.p.	↑ (n. accumbens) ↑ (n. accumbens) ↑ (n. accumbens) ↑↑ (striatum) ↑ (striatum) n. accumbens, frontal cortex)	↑↑ ↑↑ ↑↑ ↑ ↑↑	Baumann et al. (2012) Kehr et al. (2011) Suyama et al. (2016) Shortall et al. (2015) Gołembiowska et al. (2016)
Methylone 0.3, 1 mg/kg i.v.	↑ (n. accumbens)	↑↑	Baumann et al. (2012)
MDPV			
0.1, 0.3 mg/kg i.v.	↑ (n. accumbens)	No effect	Baumann et al. (2013)
4F-methcathinone 4Br-methcathinone 4Cl-methcathinone 1, 3.2, 10 mg/kg i.p.	↑ (n. accumbens) ↑ (n. accumbens) ↑↑ (n. accumbens)	↑↑ ↑↑ ↑	Suyama et al. (2016)
4-CF3-methcathinone 1, 3 mg/kg i.v.	No effect	↑ (n. accumbens)	Cozzi et al. (2013)

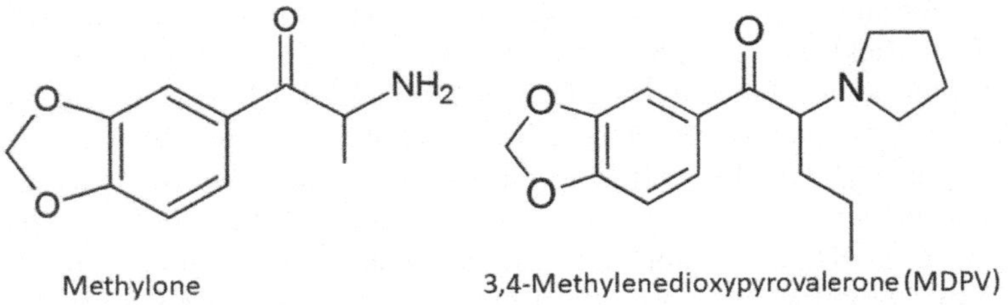

Fig. 7.1 Structural formulas of methylone and 3,4-methylenedioxypyrovalerone (MDPV)

The summary of the in vivo effects of cathinone analogs on neurotransmitter release is presented in Table 7.1.

Figures 7.1 and 7.2 present structural formulas of cathinone analogs.

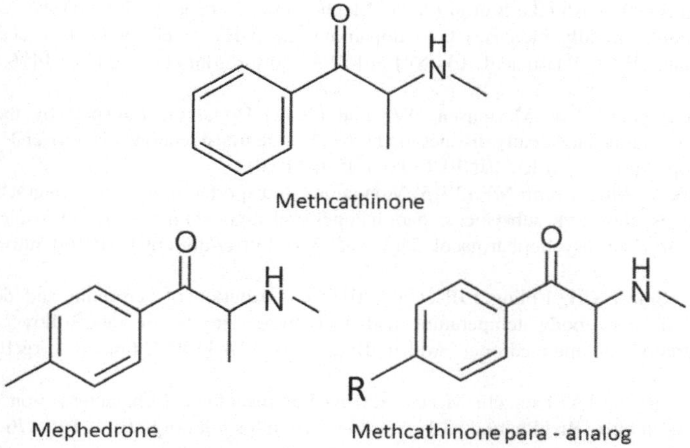

Fig. 7.2 Structural formulas of methcathinone, mephedrone, and a *para*-analog of methcathinone. R = F, Br, Cl or CF3

Acknowledgements Supported by the National Science Centre (NCN), Cracow, Poland (grant no 2013/09/B/NZ7/04104).

References

Baumann MH, Ayestas MA, Partilla JS et al (2012) The designer methcathinone analogs, mephedrone and methylone are substrates for monoamine transporters in brain tissue. Neuropsychopharmacology 37(5):1192–1203. https://doi.org/10.1038/npp.2011.304.2011.304

Baumann HM, Partilla JS, Lehner KR et al (2013) Powerful cocaine-like action of 3,4-methylenedioxypyrovalerone (MDPV), a principal constituent of psychoactive 'bath salts' product. Neuropsychopharmacology 38(4):552–562. https://doi.org/10.1038/npp.2012.204

Cameron K, Kolanos R, Verkariya R et al (2013) Mephedrone and methylenedioxypyrovalerone (MDPV), major constituents of *bath salts*, produce opposite effects at the human dopamine transporter. Psychopharmacology (Berl) 227(3):493–499. https://doi.org/10.1007/s00213-013-2967-2

Cozzi NV, Brandt SD, Daley PF et al (2013) Pharmacological examination of trifluoromethyl ring-substituted methcathinone analogs. Eur J Pharmacol 699(1–3):180–187. https://doi.org/10.1016/j.ejphar.2012.11.008

Eshleman AJ, Wolfrum KM, Hatfield MG et al (2013) Substituted methcathinones differ in transporter and receptor interactions. Biochem Pharmacol 85(12):1803–1815. https://doi.org/10.1016/j.bcp.2013.04.004

Gołembiowska K, Jurczak A, Kamińska K et al (2016) Effect of some psychoactive drugs used as 'legal highs' on brain neurotransmitters. Neurotox Res 29(3):394–407. https://doi.org/10.1007/s12640-015-9569-1

Gygi MP, Fleckenstein AE, Gibb JW, Hanson GR (1997) Role of endogenous dopamine in the neurochemical deficits induced by methcathinone. J Pharmacol Exper Ther 283(3):1350–1355

Hadlock GC, Webb KM, McFadden LM et al (2011) 4-methylmethcathinone (mephedrone): neuropharmacological effects of a designer stimulant of abuse. J Pharmacol Exp Ther 339 (2):530–536. https://doi.org/10.1124/jpet.111.184119

Kehr J, Ichinose F, Yoshitake S et al (2011) Mephedrone, compared with MDMA (ecstasy) and amphetamine, rapidly increases both dopamine and 5-HT levels in nucleus accumbens of awake rats. Br J Pharmacol 164(8):1949–1958. https://doi.org/10.1111/j.1476-5381.2011.01499.x

Larsen MB, Sonders MS, Mortensen OV et al (2011) Dopamine transport by the serotonin transporter: a mechanistically distinct mode of substrate translocation. J Neurosci 31(17):6605–6615. https://doi.org/10.1523/JNEUROSCI.0576-11.2011

Rickli A, Hoener MC, Liechti ME (2015) Monoamine transporter and receptor interaction profiles of novel psychoactive substances: para-halogenated amphetamines and pyrovalerone cathinones. Eur Neuropsychopharmacol 25(3):365–376. https://doi.org/10.1016/j.euroneuro.2014.12.012

Shortall SE, Spicer CH, Ebling FJP et al (2015) Contribution of serotonin and dopamine to changes in core body temperature and locomotor activity in rats following repeated administration of mephedrone. Addict Biol 21(6):127–1139. https://doi.org/10.1111/adb.12283

Simmler LD, Buser TA, Donzelli M et al (2013) Pharmacological characterization of designer cathinones in vitro. Br J Pharmacol 168(2):458–470. https://doi.org/10.1111/j.1476-5381.2012.02145.x

Simmler LD, Liechti ME (2017) Interactions of cathinone NPS with human transporters and receptors in transfected cells. In: Baumann MH, Glennon RA, Wiley JL (eds). Curr Top Behav Neurosci. Springer Inter Pub AG, 32:49–72. http://doi.org/10.1007/7854_2016_20

Suyama JA, Sakloth F, Kolanos R et al (2016) Abuse-related neurochemical effects of para-substituted methcathinone analogs in rats: microdialysis studies of nucleus accumbens dopamine and serotonin. J Pharmacol Exp Ther 356(1):182–190. http://doi.org/10.1124/jpet.115.229559

Wise RA (2008) Dopamine and reward: the anhedonia hypothesis 30 years on. Neurotox Res 14 (2–3):169–183. https://doi.org/10.1007/BF03033808

Chapter 8
Behavioral Profiles and Underlying Transmitters/Circuits of Cathinone-Derived Psychostimulant Drugs of Abuse

Steven J. Simmons, Erin Kim, Taylor A. Gentile, Ali Murad, John W. Muschamp and Scott M. Rawls

Abstract Plant-derived cathinone (found in khat of *C. edulis*, a shrub native to Horn of Africa and Arabian Peninsula) is structurally similar to amphetamine and produces psychosomatic, behaviorally activating effects. Alterations to one or more motifs of cathinone's molecular structure can yield novel psychoactive compounds with greater potency, toxicity, and abuse liability than the parent compound. These synthetic cathinone drugs, including 4-methylmethcathinone (mephedrone) and 3,4-methylenedioxypyrovalerone (MDPV), are manufactured in clandestine laboratories with intent for recreational use and have been the subject of intensive research on basic and preclinical levels over the past decade. For years, synthetic cathinone drugs avoided legal persecution by vending formulations in packages marked "not for human consumption." Even after scheduling by the Drug Enforcement Agency, novel formulations of synthetic cathinone drugs continue to be seized across the United States and internationally with emergency room visits and postmortem toxicology reports supporting this drug class as a persisting public health concern. This chapter will provide evaluation of studies that examine the behavioral effects following synthetic cathinone use as well as provide discussion of the transmitters and circuits that underlie these effects. Similarly to cocaine, amphetamine, and 3,4-methylenedioxymethamphetamine (MDMA), synthetic cathinone drugs work by elevating extracellular levels of monoamines by augmenting release or by blocking presynaptic uptake. In animal models, synthetic cathinone drugs produce place preference, acutely prime brain reward thresholds and are readily self-administered. Available evidence supports that the potently rewarding and reinforcing effects of certain synthetic cathinone drugs are likely

S. J. Simmons · E. Kim · T. A. Gentile · A. Murad
J. W. Muschamp · S. M. Rawls (✉)
Lewis Katz School of Medicine, Temple University,
3500 North Broad St., Philadelphia, PA 19140, USA
e-mail: srawls@temple.edu

S. J. Simmons
e-mail: steven.simmons@temple.edu

J. B. Zawilska (ed.), *Synthetic Cathinones*, Current Topics in Neurotoxicity 12,
https://doi.org/10.1007/978-3-319-78707-7_8

mediated by preferential action at dopaminergic and noradrenergic transporters, while empathogenic effects more strongly align with augmentation of serotonin transmission. As additional compounds are designed throughout the United States and elsewhere, understanding the sites of action and behavioral effects remains the goals of our laboratory and others in ultimate effort to intervene and suppress addiction propensity of harmful cathinone-derived psychostimulant drugs.

Keywords Addiction models · Intracranial self-stimulation · Place conditioning Reinforcement · Reward · Self-administration · Synthetic cathinone(s)

Abbreviations

2PMPA	2-(phosphonomethyl)-pentanedioic acid
4-FMC (flephedrone)	4-fluoromethcathinone
4-MEC	4-methylethylcathinone
4-MePPP	4-methyl-α-pyrrolidinopropiophenone
4-MMC (mephedrone)	4-methylmethcathinone
5-HT	Serotonin
BMAPN	2-(methylamino)-1-(naphthalene-2-yl) propan-1-one
DA	Dopamine
DAT	Dopamine transporter
FR	Fixed ratio
GCPII	Glutamate carboxypeptidase II
GLT1	Glutamate transporter 1
Hcrt/Ox	Hypocretin/orexin
ICSS	Intracranial self-stimulation
MACHP	2-cyclohexyl-2-(methylamino)-1-phenylethanone
MAOP	2-(methylamino)-1-phenyloctan-1-one
MDEC (ethylone)	3,4-methylenedioxyethylcathinone
MDMA	3,4-methylenedioxymethamphetamine
MDMC (methylone)	3,4-methylenedioxymethcathinone
MDPV	3,4-methylenedioxypyrovalerone
mexedrone	3-methoxy-2-(methylamino)-1-(4-methylphenyl)-propan-1-one
mGluR2/3	Metabotropic glutamate receptors 2 and 3
NA	Noradrenaline
NAAG	*N*-acetylaspartylglutamate
NAcc	Nucleus accumbens
naphyrone	Naphthylpyrovalerone
NET	Noradrenaline transporter
PIPP	α-piperidinopropiophenone
PIVT	α-piperidinopentiothiophenone
PR	Progressive ratio
SERT	Serotonin transporter
USV	Ultrasonic vocalization

VTA	Ventral tegmental area
α-PVP	α-pyrrolidinopentiophenone

8.1 Introduction

Cathinone [(2*S*)-2-amino-1-phenylpropan-1-one] is an alkaloid derived from khat, fresh leaves and twigs of the *Catha edulis* shrub, that shares structural and functional similarity to phenylisopropylamine (amphetamine) with the added feature of a ketone group at the β-carbon atom. Khat is chewed in social settings by natives of East Africa and of the Arabian Peninsula which can result in pro-social stimulatory effects including euphoria and allayed fatigue (see Chap. 3). Alterations to one or more motifs of cathinone's structure can yield psychoactive compounds possessing reinforcing effects with high abuse potential. "First-generation" cathinone-derived synthetic compounds including 4-methylmethcathinone (4-MMC, herein termed "mephedrone"), 3,4-methylenedioxymethcathinone (MDMC, herein termed "methylone"), and 3,4-methylenedioxypyrovalerone (MDPV) have led to psychosis, organ failure, and death in human abusers. Following legal status change across many countries between 2010 and 2012, "second-generation" compounds including α-pyrrolidinopentiophenone (α-PVP) and 3,4-methylenedioxyethylcathinone (MDEC, herein termed "ethylone") quickly emerged and became recognized as a worsening public health concern. A collection of case reports and preclinical laboratory studies since the 1970s has permitted an appreciable understanding of the behavioral effects elicited following cathinone-derived psychostimulant use.

This chapter will initially unfold the behavioral effects of synthetic cathinone drugs from human abusers which have largely been compiled from hospital-based case reports. Thereafter, an expansive section will chronicle current state of knowledge regarding behavioral effects of synthetic cathinone drugs from controlled laboratory animal studies. A concluding section will then describe the mechanisms (neurotransmitters and circuits) of how cathinone- and cathinone-derived synthetic compounds exert subjectively rewarding and reinforcing effects.

8.2 Psychostimulant Effects of Synthetic Cathinones in Human Abusers

Much like cocaine and amphetamine, synthetic cathinone use yields powerful although transient stimulant-like psychosomatic and behavioral effects which can include elevated heart rate, tachycardia/arrhythmia, psychosis, anxiety, agitation, self-mutilation, and induction of paranoid delusion. Certain synthetic cathinone drugs can additionally elicit an empathogenic response typified by 3,4-methylenedioxymethamphetamine (MDMA, also termed "Ecstasy" and "Molly" [readers should note prevalence of adulteration with other compounds in street formulations]). While the parent compound cathinone has a relatively slow onset of action following ingestion by khat chewing, reports find that synthetic cathinone drugs are often administered by injection or snorting to give rapid highs prior to raves/electronic dance music festivals as well as before "chemsex" gatherings. Administration of synthetic cathinone formulations is seemingly taken in effort to enhance sensory and social experiences. A series of case studies and toxicology reports highlights the high risk of adverse effects, including mortality, of synthetic cathinone use.

A case presentation of a regular mephedrone injector revealed symptoms of drug-induced psychosis in addition to appetite loss and various side effects. In this specific report, the user's increased tolerance and dependence (as interpreted from negative symptoms, including agitation, emerging during episodes of cessation) are highlighted (Bajaj et al. 2010). Similarly, in studying a group of mephedrone formulation injectors in Ireland, Van Hout and Bingham (2012) describe binge-like patterns of intravenous use (termed "slamming"; e.g., Dolengevich-Segal et al. 2016) characterized by bouts of euphoria, enhanced sensory experience (e.g., for sexual encounters), and rapid withdrawal-like "comedown" symptoms. In efforts to manage these symptoms throughout binges, including intense paranoia, violent behavior, rapid onset pain, spasms (clonus), and aggression, users reported co-administering drugs including heroin, methadone, or benzodiazepines. Physically, limb abscesses, blisters, and vein clotting were common following repeated injections in the arm, neck, or groin area. Criminality risk increased as users failed to manage negative symptoms thought drug binges. A university-conducted survey in the United Kingdom performed in 2010 makes clear the abuse potential of consecutive injections of mephedrone—of 1006 users surveyed, nearly half indicated the compound as addictive with 17.5% self-reporting dependence (Dargan et al. 2010). Some high-dose mephedrone users show signs of "serotonin syndrome" including hyperthermia and clonus which can be fatal if left untreated (e.g., Garrett and Sweeney 2010).

Similarly to psychotropic effects of mephedrone, binge insufflation of "Powdered Rush" (verified as MDPV through urine analysis) led to admission of an individual to a psychiatric ward for having experienced delusions 5 days after drug use (Antonowicz et al. 2011). A more comprehensive report on MDPV cases from two regional poison centers indicated that, out of 236 patients, 194 reported agitation and 134 reported combative behaviors with fewer counts of tachycardia, hallucinations, chest pain, and myoclonus (Spiller et al. 2011).

In 2015, the National Forensic Laboratory Information System reported the illicit compound ethylone as the most prevalent analytically confirmed seized drug within the synthetic cathinone class (9237 [47.40%] of 19,490 cases within the United States) and noted the emergence of other novel compounds including 4-fluoromethcathinone (4-FMC, also termed "flephedrone") and 3-chloromethcathinone (3-CMC). In a recent observational case, series of 305 individuals, 3-methoxy-2-(methylamino)-1-(4-methylphenyl)propan-1-one ("mexedrone") was identified as an emerging synthetic cathinone in 11 toxicologically verified patients with clinical features, including agitation, sinus tachycardia, and experience of demonic/supernatural themed hallucinations; 6 of 11 patients required physical restraint and/or sedation (Roberts et al. 2017). Clinical examination of a 31-year-old naphthylpyrovalerone ("naphyrone") injector reported restlessness and hampered vision at 5 hours post-administration as well 48-h insomnia alongside hallucinations and abnormal vital signs (Derungs et al. 2011).

8.3 Reward and Reinforcement from Cathinone-Derived Drugs: Preclinical Evidence

The following sections will overview the designs and interpretations of behavioral assays used to probe abuse liability of synthetic cathinone drugs in animal studies. Each section will initially introduce relevant assays and behavioral measures followed by a chronicle of findings from the parent compound (cathinone) and then from first- and second-generation synthetic cathinone drugs. When possible, elaboration of stereospecific effects will be provided. Select findings are summarized in Tables 8.1, 8.2, 8.3, and 8.4 which correspond to the below-detailed subsections.

8.3.1 Locomotor Activity, Behavioral Sensitization, and Drug Discrimination

Across pharmacologic classes, the effects of drugs on locomotor activity is a basic behavioral measure, but one that has special relevance to psychostimulants because they influence motor output almost by definition. While running wheels and rotometers may be deployed, most often, open-field ambulation is assessed electronically during a set time (e.g., 60–90 min) in a $\sim$40 cm^2 Plexiglas chamber with photocells arrayed near its floor. Interruption of infrared energy by a passing rodent is detected and streamed to a computer, from where it is extracted time-stamped data on location, ambulation, velocity, distance, and exploratory rearing. Given acutely, stimulant drugs typically show a non-monotonic dose–response with low doses increasing and high doses decreasing distance traveled. The descending arm of this function is explained by a concomitant increase in repetitive, tic-like stereotypy at the expense of forward locomotion. When observed in high-dose human stimulant users, this stereotypy is colloquially termed "tweaking". Another notable phenomenon is seen when stimulants are administered on alternate days over longer period. Here, locomotor effects are potentiated across testing sessions and show sensitization rather than diminution as occurs with tolerance. This amplified drug effect can be seen despite weeks of abstinence, with some studies demonstrating a single-stimulant exposure in young rats, which is sufficient to sensitize the locomotor response to an amphetamine challenge given $\geq$ 12 months later. Sensitization also corresponds to increased drug-seeking as well as

Table 8.1 Effects of cathinone and synthetic analogs on locomotor activity, behavioral sensitization, and drug discrimination

Locomotor activity, behavioral sensitization and drug discrimination						
Compound	Route	Dose(s)	Frequency	Effect	Comparator group(s)	References
CATH	i.p.	–	Acute	↑ Ambulation	Veh	Knoll (1979)
CATH	i.p.	3–20	Acute	↑ Ambulation[a]	Veh	Glennon et al. (1987)
Methcathinone		1–10				
CATH		7.5–15	Repeated[b]	Substitution (AMPH 1)[a]	Veh	
Methcathinone		3–7.5				
CATH	icv	16–64 µg	Acute	↑ Ambulation	Veh	Calcagnetti and Schechter (1992)
	NAcc	20 µg				
	SN			No effect		
S-meth-cathinone	i.p.	0.60	Acute	↑ Ambulation	R-meth-cathinone	Glennon et al. (1995)
		0.18[c]	Repeated[b]	Substitution (COC 8)	Veh	
R-meth-cathinone		0.50[c]				
CATH	p.o.	1.5	Acute	↑ Ambulation	Veh	Banjaw and Schmidt (2005)
C. edulis		200		No effect		
CATH		1.5	1x/day, 9 day	Sensitization	Challenge versus Day 1	
C. edulis		200				
MEPH	i.p.	1–3	Acute	↑ Ambulation[a]	Veh	Kehr et al. (2011)
				↓ Ambulation[a]	AMPH	
MEPH	i.p.	5–25 mg/kg	Acute	↑ Ambulation	Veh	Lopez-Arnau et al. (2012)
Methylone						
Butylone						
MDPV	i.p.	1–17	Acute	↑ Ambulation	Veh	Marusich et al. (2012)
MEPH		3–30				
Methylone		10–30				

(continued)

Table 8.1 (continued)

Locomotor activity, behavioral sensitization and drug discrimination						
Compound	Route	Dose(s)	Frequency	Effect	Comparator group(s)	References
Methedrone						
3-FMC		3–30				
4-FMC		17–30				
MDPV	i.p.	0.3–1	Acute	↑ Ambulation	Veh	Baumann et al. (2013)
					COC	
MEPH	i.p.	Variable dosing	1x/day, 7 day	Sensitization	Challenge versus Day 1	Gregg et al. (2013)
MEPH	i.p.	15–30	Acute	↑ Ambulation[a]	Veh	Motbey et al. (2013)
CATH	i.p.	1–4	Acute	↑ Ambulation	Veh	Shortall (2013)
MEPH		1–4		No effect		
		10		↑ Ambulation		
CATH		1–4	1 x/day, 2 day/week, 3 weeks	Sensitization	Challenge versus Day 1	
MEPH		1–4		No effect		
		10		Sensitization		
α-PBP	i.p.	2.5–10	Acute	↑ Ambulation	Veh	Gatch et al. (2015a)
α-PVP						
4'-MePPP		30				
α-PBP		10[a]	Repeated[b]	Substitution (COC 10)		
α-PVP						
4'-MePPP		1–50		No effect		
Methcathinone	i.p.	1–30	Acute	↑ Ambulation	Veh	Gatch et al. (2015b)
Pentedrone		5–25				
3-FMC		3–10				
Pentylone		10–30				
4-MEC		30–100				

(continued)

Table 8.1 (continued)

Compound	Route	Dose(s)	Frequency	Effect	Comparator group(s)	References
Locomotor activity, behavioral sensitization and drug discrimination						
Methcathinone		0.52[c]	Repeated[b]	Substitution (COC 10)		
Pentedrone		2.29[c]				
3-FMC		0.81[c]				
Pentylone		3.14[c]				
4-MEC		12.54[c]				
R-MEPH	i.p.	20–30	Acute	↑ Ambulation	*S*-MEPH	Gregg et al. (2015)
		Variable dosing	1x/day, 7 day	Sensitization	Challenge versus Day 1	
α-PVP	i.p.	0.7[c]	Repeated[b]	Substitution (AMPH 1)	Veh	Naylor et al. (2015)
4-MePPP		4.8[c]				
MDPV	i.p.	0.5	1x/day, 7 day	Sensitization	Challenge versus Day 1	Berquist II et al. (2016)
					Acute (COC 5)	
MDPV	i.p.	Variable dosing	1x/day, 7 day	Sensitization	Challenge versus Day 1	Gregg et al. (2016)
α-PVP	Vapor	24–240 mg/mL	Acute	↑ Ambulation	Veh	Maruisch et al. (2016)
	i.p.	1–10				
MDPV	Vapor	25–100 mg/mL	Acute	↑ Ambulation	Veh	Nguyen et al. (2016)
MEPH		100–200 mg/mL		No effect		
MDPV	i.p.	1–3	Repeated[b]	Substitution (MEPH 1)	Veh	Berquist II et al. (2017)
PIPP	i.p.	1–30	Acute	No effect	Veh	Botanas et al. (2017c)
PIVT		1–4				

(continued)

Table 8.1 (continued)

Locomotor activity, behavioral sensitization and drug discrimination						
Compound	Route	Dose(s)	Frequency	Effect	Comparator group(s)	References
BMAPN	i.p.	30	Acute	↑ Ambulation	Veh	Botanas et al. (2017b)
			1x/day, 7 day	Sensitization		
MACHP	i.p.	30	Acute	↑ Ambulation	Veh	Botanas et al. (2017a)
MAOP						
α-PVT	i.p.	5.2[c]	Repeated[b]	Substitution (COC 5.6)	Veh	Cheong et al. (2017)
		2.0[c]		Substitution (METH 0.3)		

Doses are in mg kg^{-1} unless otherwise indicated. Values in parentheses indicate dose in mg kg^{-1}. A symbol of "–" indicates a value not found within the corresponding reference

[a]Either effect did not reach statistical significance (trend was observed) or specific comparisons against comparator group not performed

[b]Repeated until criterion is reached (typically, >80% discrimination on drug-paired lever)

[c]Dose listed is effective for 50% substitution (i.e., ED_{50})

AMPH amphetamine, *BMAPN* 2-(methylamino)-1-(naphthalene-2-yl) propan-1-one, *CATH* cathinone, *COC* cocaine, *FMC* fluoromethcathinone, *icv* intracerebroventricular, *i.p.* intraperitoneal, *iv* intravenous, *MACHP* 2-cyclohexyl-2-(methylamino)-1-phenylethanone, *MAOP* 2-(methylamino)-1-phenyloctan-1-one, *MDPV* 3,4-methylenedioxypyrovalerone, *MEC* 4-methylethylcathinone, *MEPH* mephedrone, *MePPP* 4-methyl-α-pyrrolidinopropiophenone, *METH* methamphetamine, NAcc nucleus accumbens, *PIPP* α-piperidinopropiophenone, *PIVT* α-piperidinopentiothiophenone; *α-PVT* α-pyrrolidinopentiophenone, *po* oral, *SN* substantia nigra, *Veh* vehicle

Table 8.2 Effects of cathinone and synthetic analogs in the place conditioning assay

Compound	Route	Dose(s)	Frequency	Effect	Comparator group(s)	References
Place conditioning						
CATH	i.p.	0.8	8 day (4 day/side)	Preference	Veh-paired side	Schechter (1991)
α-PBP	i.p.	3–10	2 day (2x/side/day)	Preference	Veh-paired side	Gatch et al. (2015a)
α-PVP		0.3–10				
4'-MePPP		3–100		No effect		
R-MEPH	i.p.	30	8 day (4 day/side)	Preference	Veh-paired side	Gregg et al. (2015)
S-MEPH		5–30		No effect		
MDPV	i.p.	2	8 day (4 day/side)	Preference	Veh-paired side	Gregg et al. (2016)
α-PVP	Vapor	72 mg/mL	3 day (2x/side/day)	Preference	Veh-paired side	Maruisch et al. 2016
MACHP	i.p.	10–30	8 day (1x/day, 4 day/side)	Preference	Veh-paired side	Botanas et al. (2017a)
MAOP						
BMAPN	i.p.	10–30	8 day (1x/day, 4 day/side)	Preference	Veh-paired side	Botanas et al. (2017b)
PIPP	i.p.	10–30	8 day (1x/day, 4 day/side)	Preference	Veh-paired side	Botanas et al. (2017c)
PIVT		3–30		No effect		
α-PVT	i.p.	10–50	8 day (1x/day, 4 day/side)	Preference	Veh-paired side	Cheong et al. (2017)

Doses are in mg kg^{-1} unless otherwise indicated. Values in parentheses indicate dose in mg kg^{-1}

AMPH amphetamine, *BMAPN* 2-(methylamino)-1-(naphthalene-2-yl) propan-1-one, *CATH* cathinone, *COC* cocaine; *i.p.* intraperitoneal, *iv* intravenous, *MACHP* 2-cyclohexyl-2-(methylamino)-1-phenylethanone, *MAOP* 2-(methylamino)-1-phenyloctan-1-one, *MDPV* 3,4-methylenedioxypyrovalerone, *MEPH* mephedrone, *MePPP* 4-methyl-α-pyrrolidinopropiophenone, *METH* methamphetamine, *PIPP* α-piperidinopropiophenone, *PIVT* α-piperidinopentiothiophenone, *α-PVT* α-pyrrolidinopentiophenone, *Veh* vehicle

Table 8.3 Self-administration of cathinone and synthetic analogs

Self-Administration						
Compound	Route	Dose(s)	Frequency	Effect	Comparator group(s)	References
CATH	iv	–	3 h/day (FR10), 5 day	↑ Infusions	COC	Knoll (1979)
					AMPH	
MEPH	iv	0.70	4 h/day (FR1), 8 day	↑ Infusions	Veh	Hadlock et al. (2011)
Methylone	iv	0.05–0.5	2 h/day (FR1), 21 day	↑ Active presses	Inactive presses	Watterson et al. (2012a)
MDPV	iv	0.05–0.2	2 h/day (FR1), 10 day	↑ Active presses	Inactive presses	Watterson et al. (2012b)
MDPV	iv	0.01	2 h/day (FR5), 10 day	↑ Infusions	METH (0.01)	Aarde et al. (2013a)
		0.10	~3 h/d (PR)	↑ Breakpoint	METH (0.10)	
MEPH	iv	1.5	~3 h/day (PR), ≥ 6 day	↑ Breakpoint	Veh	Aarde et al. (2013b)
MEPH	iv	0.1	2 h/day (FR1), 10 day	↑ Infusions	METH (0.03)	Motbey et al. 2013
		0.3			METH (0.1)	
MEPH	iv	0.5	2 h/day (FR1), 14 day	↑ Infusions	MDMA (0.5)	Creehan et al. (2015)
Methylone						
MACHP	iv	0.1–0.3	2 h/day (FR1), 7 day	No effect	Veh	Botanas et al. (2017a)
		1.0		↑ Infusions		
MAOP		0.1–3		No effect		
BMAPN	iv	0.1, 1	2 h/day (FR1), 7 day	No effect	Veh	Botanas et al. (2017b)
		0.3		↑ Infusions		
PIPP	iv	0.1–1	2 h/day (FR1), 7 day	No effect	Veh	Botanas et al. (2017c)
PIVT						
α-PVT	iv	0.1–1	2 h/day (FR1), 10 day	↑ Infusions	Veh	Cheong et al. (2017)
α-PVP	iv	0.006–0.025	2 h/day (FR5), ≥3 day	↑ Infusions	Veh	Huskinson et al. (2017)
4-MePPP		0.2–0.4				
4-MEC		0.8				

Doses are in mg kg^{-1} unless otherwise indicated. Values in parentheses indicate dose in mg kg^{-1}. A symbol of "–" indicates a value not found within the corresponding reference
AMPH amphetamine, *BMAPN* 2-(methylamino)-1-(naphthalene-2-yl) propan-1-one, *CATH* cathinone, *COC* cocaine, *i.p.* intraperitoneal, *iv* intravenous, *MACHP* 2-cyclohexyl-2-(methylamino)-1-phenylethanone, *MAOP* 2-(methylamino)-1-phenyloctan-1-one, *MDPV* 3,4-methylenedioxypyrovalerone, *MEPH* mephedrone, *MePPP* 4-methyl-α-pyrrolidinopropiophenone, *METH* methamphetamine, *PIPP* α-piperidinopropiophenone, *PIVT* α-piperidinopentiothiophenone, *α-PVT* α-pyrrolidinopentiophenone, *Veh* vehicle

Table 8.4 Effects of cathinone and synthetic analogs on brain reward thresholds in the intracranial self-stimulation (ICSS) assay

Intracranial Self-Stimulation						
Compound	Route	Dose(s)	Frequency	Effect	Comparator group(s)	References
MEPH	i.p.	10	Acute	↑ Stimulation	Baseline (Veh)	Robinson et al. (2012)
					COC	
Methylone	i.p.	1–10	Acute	↓ Threshold[a]	Baseline (Veh)	Watterson et al. (2012a)
MDPV	i.p.	0.1–2.0	Acute	↓ Threshold	Baseline (Veh)	Watterson et al. (2012b)
Methcathinone	i.p.	0.32–1.0	Acute	↑ Stimulation[a]	Baseline (Veh)	Bonano et al. (2014)
MDPV		0.32–3.2				
Methylone		3.2–10.0				
α-PVP	i.p.	0.3–1.0	Acute	↓ Threshold	Baseline (Veh)	Watterson et al. (2015)
4-MEC		10–30				
R-MEPH	i.p.	1.0–3.2	Acute	↑ Stimulation	Baseline (Veh)	Gregg et al. (2015)
S-MEPH						
R-CATH	i.p.	3.2–10	Acute	↑ Stimulation	Baseline (Veh)	Hutsell et al. (2016)
S-CATH		0.32–1.0				
R-MEPH		3.2				
S-MEPH		1.0–3.2		↓ Stimulation		
MDPV	Vapor	100 mg/mL	Acute	↓ Threshold	Baseline (Veh)	Nguyen et al. (2016)
MEPH		200 mg/mL				

Doses are in mg kg^{-1} unless otherwise indicated. Values in parentheses indicate dose in mg kg^{-1}

[a]Either effect did not reach statistical significance (trend was observed) or specific comparisons against comparator group not performed

AMPH amphetamine, *BMAPN* 2-(methylamino)-1-(naphthalene-2-yl) propan-1-one, *CATH* cathinone; *COC* cocaine; *i.p.* intraperitoneal, *iv* intravenous, *MDPV* 3,4-methylenedioxypyrovalerone, *MEC* 4-methylethylcathinone, *MEPH* mephedrone, *METH* methamphetamine, *α-PVT* α-pyrrolidinopentiophenone, *Veh* vehicle

drug-mediated neuroadaptations in gene expression, transmitter release, and cell morphology. Together, these findings have led to the hypothesis that sensitization represents a behavioral endophenotype with relevance to human drug addiction.

In contrast, the task of identifying and distinguishing pharmacological mechanisms ab initio has been assisted by drug discrimination studies. This paradigm hinges on the use of the drug state itself as an interoceptive cue, much like an externally presented light, tone, or odor may predict drug availability in a more traditional operant learning task. Accordingly, the extent to which a test drug is perceived as similar to a food-paired training drug is reflected in operant responses elicited by varied doses of the test drug. For example, the capacity of cocaine to substitute for amphetamine is high, to caffeine lower, and to mescaline lower still or absent. The phenomenon of robust stimulus generalization between drugs with potentially shared mechanisms is one common test used to predict abuse liability of candidate compounds before entry into human clinical trials.

Contained in a 1979 issue of *NIDA Research Monographs*, Knoll describes how systemic cathinone produces comparable elevations in ambulatory activity in mice compared to amphetamine (Knoll 1979). Years later, cathinone's hyperlocomotor effect was revealed to be centrally mediated as intracerebroventricular injection produced elevations in ambulation at comparable rates as systemic administration (Calcagnetti and Schechter 1992). Behavioral sensitization is produced from daily oral administration of either cathinone (1.5 mg/kg) or *C. edulis* extract (200 mg/kg) across a 9-day administration regimen (Banjaw and Schmidt 2005). In studying structural analogs, monomethylated cathinone ("methcathinone") possesses greater potency relative to cathinone in elevating locomotor activity and additionally substitutes for amphetamine in a drug discrimination assay (Glennon et al. 1987). Upon investigation of enantiomeric potencies, *S*-methcathinone elicits greater elevations in locomotor activity as well as ability to substitute for either cocaine or *S*-amphetamine relative to its racemate (Glennon et al. 1995).

The first-generation synthetic cathinone drug mephedrone elevates locomotor activity in adolescent rats upon systemic injection (15 or 30 mg/kg, i.p.) although effects tended to be weaker than amphetamine-evoked hyperlocomotion (Kehr et al. 2011; Motbey et al. 2013). In efforts to mimic weekend drug use as reported from some users (injections 1x/day, 2x/week across 3 weeks), Shortall and colleagues (2013) observed a dose-dependent sensitized hyperlocomotor response on the last injection of mephedrone (10 mg/kg/inj, i.p.) as well as to the parent compound cathinone (4 mg/kg/inj, i.p.) relative to ambulatory activity measured after initial injections of either compound. Similarly, racemic mephedrone produces behavioral sensitization of repetitive horizontal movements, movements of presumably stationary test subjects breaking a single photobeam within the activity-monitoring apparatus, using a variable-dose injection regimen (Gregg et al. 2013). Enantiomeric effects of mephedrone were elucidated by Gregg and colleagues (2015) who found that *R*-mephedrone more robustly elevates locomotor activity and induces a sensitized locomotor response after repeated injections whereas its racemate did not induce behavioral sensitization. In a comprehensive comparison study of several first-generation synthetic cathinone drugs, injection of methylone or butylone produced more persistent elevations in locomotor activity in mice relative to mephedrone (which itself evoked significant although transient elevations in locomotor activity relative to vehicle-treated control subjects) (López-Arnau et al. 2012). All three compounds were, however, noted to induce their peak effect shortly after injection, suggesting a shared rapid onset property throughout this class of first-generation synthetic cathinone drugs. Similarly, MDPV acutely elevates ambulatory activity in rats and mice (Aarde et al. 2013a; Baumann et al. 2013; Marusich et al. 2012) and induces locomotor sensitization by itself or when combined with moderate-dose mephedrone (2 mg/kg) across a 7-day injection regimen (Berquist et al. 2016).

In drug discrimination against methamphetamine, the second-generation cathinone-derived psychostimulant drugs α-pyrrolidinopentiophenone (α-PVP; also known as "flakka") and 4-methyl-α-pyrrolidinopropiophenone (4-MePPP) showed full substitution ability (Gatch et al. 2015a; Naylor et al. 2015). 4-Methylethylcathinone (4-MEC) additionally substituted for amphetamine but at significantly higher doses, suggesting minimal potency relative to others within the second-generation class (Gatch et al. 2015b). A separate study found that α-pyrrolidinopentiothiophenone (α-PVT), an analog of α-PVP with phenyl ring substituted with thiophene, fully substituted for both cocaine and amphetamine (Cheong et al. 2017). Upon comparing systemic injection versus vapor inhalation routes of administration, Marusich and colleagues (2016) demonstrated comparable hyperlocomotor effects of α-PVP in both male and female mice from both administration methods. The authors promote the observation that inhalation of illicit drugs including methamphetamine and cathinone-derived psychostimulants—given the compounds are easily volatilized—yields subjective effects with more rapid onset relative to alternative administration routes (e.g., nasal inhalation [snorting]). Comparable hyperlocomotor effects were observed following the inhalation of MDPV as well as mephedrone using an electronic cigarette vaporization technology (Nguyen et al. 2016). Modification of cathinone's α-carbon with cyclohexyl and n-hexyl substituents yielded novel compounds 2-cyclohexyl-2-(methylamino)-1-phenylethanone (MACHP) and 2-(methylamino)-1-phenyloctan-1-one (MAOP), respectively, and both drugs produced locomotor sensitization at high doses (30 mg/kg) but at noticeably reduced ambulation compared rats receiving daily methamphetamine (1 mg/kg) injections (Botanas et al. 2017a). Similarly, naphthalene substitution in the aromatic ring yielded 2-(methylamino)-1-(naphthalene-2-yl) propan-1-one (BMAPN) which produced locomotor sensitization across a 7-day injection regimen (Botanas et al. 2017b).

Effects of cathinones on locomotor activity, behavioral sensitization, and drug discrimination are summarized in Table 8.1.

8.3.2 Place Conditioning as a Measure of Passive Reward-Context Association

Place conditioning is an assay used by addiction researchers that involves a test subject associating a distinct environment/context (e.g., via visual, tactile, and/or olfactory cues) with an interoceptive state following administration of an agent such as cocaine or amphetamine. A two- or three-chamber apparatus is used with each chamber possessing a unique set of environmental/contextual cues as indicated above. Over repeated training sessions whereby the test subject is confined to one environment/context of the apparatus, the test subject begins associating the subjective effects of an agent with the unique environment/context in which the test subject is confined within. The alternative environment/context is paired with a vehicle injection as a control condition. After a series of drug-context pairings, the test subject is allowed to freely shuttle between both sides of the apparatus, and time on each environment/context is measured. Unsurprisingly, reinforcing psychostimulant drugs such as cocaine and amphetamine induce place preference in rats and mice as indicated by greater time spent on the psychostimulant-paired side of the apparatus during the test session. For detailed discussion on this assay, readers are referred to Tzschentke (1998).

An 8-day systemic cathinone (0.8 mg/kg, i.p) injection regimen (cathinone 4 days, vehicle 4 days) produced significant place preference in rats trained previously to discriminate cathinone from vehicle using an operant procedure (Schechter 1991). The report additionally indicates that conditioning with a higher dose of cathinone (1.2 mg/kg, i.p.) produced a significant place preference in rats without prior operant training. First-generation synthetic cathinones mephedrone, methylone, and MDPV produced significant place preference in mice with greatest potency and widest range of effective doses observed from animals receiving MDPV (Karlsson et al. 2014). In an invertebrate environmental place conditioning assay, racemic mephedrone produced significant preference with greater potency observed following *R*-mephedrone exposure relative to its racemate (Vouga et al. 2015). Comparatively, *R*-mephedrone dose-dependently produced place preference in rats at comparable preference levels compared to the racemic mixture, whereas *S*-mephedrone failed to produce detectable place preference (Gregg et al. 2015).

Of the second-generation synthetic cathinone drugs, α-PVP elicited significant place preference whereas 4-MePPP did not (Gatch et al. 2015a), an effect potentially due to short duration of effect following a single injection. Nelson and colleagues (2017) showed that, in addition to place preference across a wide range of doses, α-PVP reduced saccharin consumption in a conditioned taste aversion assay which supports that α-PVP possesses both rewarding and aversive effects not uncommon among psychostimulant drugs. Other studies found that α-PVT dose-dependently elicited place preference in mice (Cheong et al. 2017). Upon modification of cathinone's nitrogen atom with piperidine ring and in effort to predict new compounds of potential abuse, Botanas and others (2017c) reported a significant dose-dependent place preference of laboratory-synthesized α-piperidinopropiophenone (PIPP; analog of methcathinone) but not of α-piperidinopentiothiophenone (PIVT; analog of α-PVT). Both MACHP and MAOP produced place preference at moderate (10 mg/kg, i.p.) and high (30 mg/kg, i.p.) doses in rats as did BMAPN across the same doses (Botanas et al. 2017a, b).

Effects of cathinones on place preference are summarized in Table 8.2.

8.3.3 Intravenous Drug Self-administration Captures Volitional Aspect of Drug Taking

Intravenous drug self-administration captures the volitional aspect of drug taking as observed in humans from nonhuman animal subjects under controlled laboratory settings. In brief, this assay allows for assessment of reinforcing efficacy of a compound by measuring operant responses (e.g., lever pressing and wheel spinning) performed by test subjects in effort to retrieve intravenous drug infusions.

Schuster and Johanson (1979) initially observed rates of operant responding for intravenous cathinone delivery that surpassed levels reached for cocaine and amphetamine when nonhuman primates were trained to press ten times for each infusion (referred to as a fixed-ratio 10 [FR10] schedule of reinforcement). When provided opportunity for an equipotent dose of intravenous cocaine, nonhuman primates were found to respond at appreciable rates for an intravenous cathinone option (35–40% of responses; Woolverton and Johanson 1984). Cathinone was later found to be self-administered by rats with approximately twofold greater potency relative to self-administered cocaine (Gosnell et al. 1996).

Of the first-generation synthetic cathinone drugs, mephedrone (~0.7 mg/kg/inf) was shown to be self-administered by rats with a sustained increase in responding during days 5–8 of daily self-administration (Hadlock et al. 2011). These results were supported by later studies showing significant responding for mephedrone with greatest effort observed from rats trained to self-administer a high dose (~1 mg/kg/inf) (Aarde et al. 2013b; Motbey et al. 2013). Similar effects were observed when female rats were permitted opportunity to self-administer mephedrone (Creehan et al. 2015). Among the initial reports documenting abuse liability of MDPV, Watterson and colleagues (2012b) revealed potently reinforcing effects of intravenous MDPV from self-administering rats. Further, at the low dose (0.05 mg/kg/inf) tested, rats showed notably high responding on the first day of self-administration access followed by a transient decrease and eventual return to ~60 infusions per 2-h sessions. During 6-h extended access testing conditions, 0.1 and 0.2 mg/kg/inf doses of self-administered MDPV showed patterns of escalated responding over time which is considered by many as a principal feature of transition from episodic to compulsive drug taking (e.g., Ahmed and Koob 1998). It was then found that rats show greater motivation for MDPV compared to methamphetamine when tested under progressive ratio (PR) access conditions (Aarde et al. 2013a) and comparable rates of responding at one-tenth dose compared to intravenous cocaine under FR1 access conditions (Simmons et al. 2016). Across all doses tested, intravenous methylone was shown to be readily self-administered with increased responding occurring after 4+ days of self-administration but failed to show an escalation in intake when permitted extended access (Watterson et al. 2012a); relatively weak reinforcing effects of intravenous methylone were also shown during self-administration in female rats (Creehan et al. 2015). Not surprisingly, rats exerted greatest effort to retrieve high-dose methylone (~ 0.5 mg/kg/inf) when given either 2- or 6-h access.

Relative to methamphetamine, the second-generation synthetic cathinone α-PVP demonstrated greater potency during intravenous self-administration in rats with lowest yet stably reinforcing dose established at 0.012 mg/kg/inf (Huskinson et al. 2017). By assessing economic demand, Huskinson and colleagues (2017) further showed that α-PVP and 4-MePPP but not 4-MEC are worked for with greater effort in rats relative to responding for methamphetamine. α-PVT produced similar reinforcing efficacy under FR and PR schedules of reinforcement (Cheong et al. 2017). Curiously, while rats were more motivated to retrieve high-dose α-PVT as evaluated by higher breakpoints under PR testing conditions, moderate-dose α-PVT (~ 0.3 mg/kg/inf) was responded for in greatest quantity under FR access conditions. Neither the α-PVT analog PIVT nor methcathinone analog PIPP was self-administered by rats despite a modest place preference induced by systemic PIPP (Botanas et al. 2017c). MACHP was self-administered by rats at 1 mg/kg/inf but not at lower doses suggesting abuse potential but low reinforcing potency (Botanas et al. 2017a). Comparatively, BMAPN sustained responding at 0.3 mg/kg/inf but not at lower or higher doses tested (Botanas et al. 2017a).

Effects of cathinones on intravenous drug self-administration are summarized in Table 8.3.

8.3.4 Intracranial Self-stimulation (ICSS) of Reward Pathways

Electrical stimulation to certain brain structures is readily responded for in test subjects. The intensity of electrical stimulation is oftentimes a significant determining factor as to whether an animal will perform operant responses for stimulation. Acute administration of drugs of abuse has been shown to "prime" brain reward function and in turn decrease the stimulation frequency needed to support operant responding. Conversely, spontaneous or precipitated withdrawal from chronic drug administration can decrease brain reward function as interpreted from increases in stimulation frequency needed to support operant responding.

Methcathinone, MDPV, and methylone (in descending order of potency) but not mephedrone acutely facilitate electrical self-stimulation in rats as evaluated by greater responding for a single frequency of electrical stimulation (Bonano et al. 2014; Watterson et al. 2012a, b). In contrast, mephedrone was shown to induce a strong leftward shift of rate-frequency function—interpreted as potentiating brain reward function—when mice responded for incrementally lower electrical stimulation frequencies (Robinson et al. 2012). Methylone was found to modestly suppress thresholds needed to support self-stimulation in rats upon acute injection of doses ranging from 1 to 10 mg/kg (Watterson et al. 2012a). While both isomers of the parent compound cathinone acutely increase responding for electrical intracranial stimulation, only *R*-mephedrone was found to support brain stimulation reward (Hutsell et al. 2016). In fact, *S*-mephedrone dose-dependently suppressed operant responding for intracranial stimulation. This finding supported earlier work demonstrating greater facilitation of operant responding for intracranial stimulation by acute injection of *R*- versus *S*-mephedrone (Gregg et al. 2015). Similarly to MDPV and methylone, second-generation synthetic cathinone α-PVP and 4-MEC dose-dependently decreased operant responding for self-stimulation in discrete trials current threshold procedure with greater potency observed from α-PVP (Watterson et al. 2015).

Effects of cathinones on electrical intracranial self-stimulation are summarized in Table 8.4.

8.3.5 Other Behavioral Effects of Synthetic Cathinones in Animals Mapped Against User Accounts

Systemic cathinone can additionally affect sexual behavior in male rats which supports anecdotal reports of enhanced libido following khat chewing in human users. Following a 15-day injection regimen (5 mg/kg/d, i.p.), cathinone significantly reduced mount and intromission latencies toward a sexually receptive female relative to latencies from vehicle-injected control rats (Taha et al. 1995). Chronic cathinone additionally increased anogenital sniffing of receptive female and increased post-ejaculatory intervals. In a follow-up study, *C. edulis* extract (including cathinone but also other endogenous compounds) was shown to dose-dependently affect sexual motivation and performance (Mohammed and Engidawork 2011). Specifically, mild dosing (100 mg/kg, daily for 4 weeks) decreased mount and intromission latencies to sexually receptive females, doubled serum testosterone level, and halved epididymal sperm count. At higher doses, behavioral measures revealed impaired sexual motivation and performance, decrements in testosterone and sperm count, and an elevation in serum cortisol. Of 116 nonspecific "bath salt" users recruited in an online survey, 42% reported elevated sex most if not all of the time following preferred compound administration (Johnson and Johnson 2014). From preclinical evidence and anecdotal reporting, chronic dosing of cathinone-derived drugs including khat or synthetic compounds may enhance sex drive through perhaps enhanced sensory experience during sexual encounters but are not at present the subject of study for medical management of sexual disorders.

An ethologically relevant measure proposed to reflect situational valence (positive versus negative) from laboratory rats has been through recording and analysis of ultrasonic vocalizations (USVs) (e.g., Knutson et al. 2002). Briefly, while lower pitched USVs around 22 kHz are emitted in negatively valenced situations, including predatory odor exposure and following electrical footshock, positively valenced 50-kHz USVs (belonging to "frequency-modulated" and "trill"-like call-type categories) are elicited during positively valenced situations, including copulation, reinforced intracranial electrical stimulation as well as during intravenous self-administration of psychostimulant drugs including cocaine (Barker et al. 2014; Burgdorf et al. 2000; Ma et al. 2010). Recently, Simmons and colleagues (2016) observed that intravenous MDPV ($\sim$0.06 mg/kg/inf) elicited robust rates of 50-kHz USVs during the drug "loading" period (initial $\sim$10 infusions) as well as during a 5-min anticipatory period in which the subject was placed in the environment paired with MDPV self-administration. 50-kHz USVs following drug administration map tightly onto periods in which users report feeling "high", and thus this measure may support the ability to interpret "likeability" of novel drugs in a manner resembling human self-reports in animal laboratory settings.

Complimenting erratic behaviors observed from human synthetic cathinone users, repeated oral administration of either cathinone or *C. edulis* extract enhanced basal aggressive behavior in male rats relative to vehicle-treated control subjects (Banjaw et al. 2006). Upon conducting a functional observational battery, all tested synthetic cathinone drugs (including first-generation compounds MDPV, mephedrone, methylone, as well as second-generation compounds 3- and 4-fluoromethcathinone [3- and 4-FMC, respectively]) produced head circling/weaving movements in mice and induced a hyperactive response (defined as fast movement of entire body) (Marusich et al. 2012). Similarly to MDPV and methamphetamine, high-dose 3-FMC (30 mg/kg, i.p.) significantly increased the frequency of stereotyped compulsive movements. In fact, 3-FMC was reported as the compound producing the greatest "behavioral burden" in the functional observational battery. Convulsions and tremors were only observed following high-dose methylone (56 mg/kg, i.p.).

Finally, impaired cognitive ability (as interpreted by significantly decreased discrimination ratios following novel object discrimination testing) was shown following chronic dosing of either high-dose racemic mephedrone (10 mg/kg/injection, 1 injection/day, 2 injections/week across 3 weeks) or MDMA (identical dosing regimen) (Shortall et al. 2013). The authors further report that chronic high-dose mephedrone caused a significant reduction of 3,4-dihydroxyphenylacetic acid (DOPAC; a principal metabolite of dopamine, DA) in hippocampal lysates when assayed 7 days following final injections; curiously, hippocampal DOPAC levels were significantly elevated from rats receiving 4 mg/kg/inj dosing of racemic mephedrone, but an association with cognitive performance was not shown.

8.4 Transmitter Systems and Circuits Underlying Behavioral Effects of Synthetic Cathinones

A ubiquitous feature of psychostimulant drugs is their ability to augment catecholamine levels within brain structures known to process rewards and emotion. Classically, augmentation of DA within nucleus accumbens (NAcc), which itself resides within ventral striatum and receives input from DA-producing afferents of ventral tegmental area (VTA), maps onto a compound's ability to enhance locomotor activity, be self-administered, and to facilitate intracranial stimulation thresholds. Empathogenic properties of compounds including MDMA tend to map onto elevations in serotonin (5-HT). Numerous studies have amassed to test the degree to which synthetic cathinone drugs influence catecholamine levels and, notably, the mechanisms through which synthetic cathinone drugs augment catecholamine neurochemistry.

8.4.1 Cathinone

In initial tests, cathinone was found to facilitate neuromuscular noradrenergic transmission *ex vivo*, an effect blocked by co-application with desipramine, the inhibitor of noradrenaline transporter NET (Knoll 1979). Soon thereafter, Kalix (1980, 1981) demonstrated cathinone's ability to stimulate release of radiolabeled DA in rabbit striatal tissues at comparable levels as amphetamine application. It was later reported that cathinone can elevate extracellular DA levels in vivo within both dorsal and ventral striatum (Pehek et al. 1990). Later work supported the idea that cathinone blocks presynaptic DA uptake as direct bilateral injection into NAcc, which receives inputs from DA-producing afferents of VTA, produced significant behavioral changes relative to vehicle-injected control subjects (e.g., on ambulation) (Calcagnetti and Schechter 1992). Further still, pharmacological intracerebroventricular pretreatment with a dopamine release inhibitor (CGS 10746B) effectively reduced cathinone place preference as well as cathinone-stimulated ambulation in rats (Calcagnetti and Schechter 1993). In sum, cathinone itself tends to elevate levels of DA and noradrenaline (NA) through release-enhancing and transport-blocking mechanisms with negligible impact on 5-HT neurochemistry.

8.4.2 First-Generation Synthetic Cathinones: Mephedrone, Methylone, and MDPV

Similarly to its parent compound cathinone, mephedrone is pharmacologically classified as a monoamine releasing agent due to the induction of presynaptic transmitter release into the synapse—this effect synergizes with an ability to bind to and block DA and 5-HT transporters (DAT and SERT, respectively). In studying stereospecific effects of mephedrone, Gregg and colleagues (2015) revealed a stronger effect of *R*-mephedrone on elevating DA release relative to its racemate while exerting negligible effect on 5-HT release.

In a comprehensive in vivo microdialysis study, Kehr and colleagues (2011) reported a ~400% increase in accumbal DA levels at 20 min following racemic mephedrone injection (4 mg/kg, s.c.), an effect that peaks at 40 min (~500%) and remains elevated with eventual return to baseline following 180 min. Effects on accumbal 5-HT levels reveal a powerful although transient effect with peak concentration measured at 20 min post-injection (~900% of baseline 5-HT) and return to baseline levels following 60 min. From cumulative assessment across 3-h post-injection, racemic mephedrone preferentially elevates 5-HT over DA in vivo (1.22 5-HT : 1.00 DA), an effect that in part matches with significantly reduced transmitter metabolite levels (Kehr et al. 2011). Relatively greater influence on 5-HT neurochemistry in vivo was additionally observed following intravenous injection of methylone (Baumann et al. 2012).

Combined effects on 5-HT and DA levels as observed following racemic mephedrone administration correspond to greater behavioral activating effects (e.g., on ambulatory activity) relative to MDMA which more selectively augments 5-HT in vivo (López-Arnau et al. 2012). Interestingly, while repeated MDMA injections lead to a relatively pronounced long-term depletion of cortical and striatal 5-HT levels, repeated administration of either mephedrone or methylone did not significantly impact long-term 5-HT integrity (Baumann et al. 2012). Regarding stereospecific behavioral effects, greater reinforcing effect of *R*-mephedrone on PR self-administration relative to its racemate may be explained by preferential augmentation of DA over 5-HT (Philogene-Khalid et al. 2017b). Similarly, facilitation of intracranial self-stimulation (ICSS) has been argued to be dependent in large part upon DA release relative to 5-HT transmission (e.g., Bauer et al. 2013). Relatively weak reinforcing efficacy of methylone, for example, during ICSS may be explained by its powerful effect on elevating accumbal 5-HT over baseline levels (~500% 20 min following 1 mg/kg dose, ~1000% 20 min following 3 mg/kg dose), whereas relatively selective enhancement of DA transmission such as with MDPV underlies a more robust reinforcing behavioral phenotype (Schindler et al. 2016).

Behavioral effects following MDPV administration seem to derive from altered DA and NA neurochemistry via potent and relatively persistent DAT and NET inhibition. MDPV has greater inhibition of DA and NA reuptakes compared to cocaine, amphetamine, and mephedrone; however, MDPV does not cause significant stimulation of DA or NA release as mephedrone does. A single injection of MDPV (0.3 mg/kg, s.c.) increases NAcc DA levels by ~400% within the first 20 min, an effect surpassing cocaine at tenfold dose which elevated DA levels by ~200% at 20 min post-injection (Baumann et al. 2013). Combining effects on behavioral and neurochemical measurements, MDPV was interpreted as having approximately tenfold greater potency than cocaine. In vitro, MDPV reduces the available monoamines for transport via DAT, NET and SERT by 50% at 0.03, 0.04, and 9.30 μM, respectively (termed IC_{50} values). Racemic mephedrone comparatively yields IC_{50} measurements for the same transporters at 0.25, 3.31, and 4.64 μM, respectively (Simmler et al. 2013). Structurally, MDPV's powerful affinity against DAT has been attributed to its carbonyl group and tertiary amine motifs (including side chain) (Kolanos et al. 2013). Finally, the reinforcing and DA-augmenting effects of racemic MDPV seem to be derived in large part from its *S* enantiomer as *S*-MDPV proved to have over twofold greater potency at blocking DAT in vitro relative to its racemate (Kolanos et al. 2015).

8.4.3 Second-Generation Synthetic Cathinones: α-PVP, 4-MEC, and Others

Second-generation synthetic cathinone drugs, including α-PVP, elicit similar effects on neurochemistry. α-PVP has been demonstrated to elevate DA levels in mouse striatum for comparable time periods relative to MDPV (Kaizaki et al. 2014). 4-MEC inhibits DAT but also elevates 5-HT via SERT binding (Iversen et al. 2013; Saha et al. 2015; Simmler et al. 2014). Using a cell line in vitro study, Eshleman and colleagues (2017) reported that 4-MEC as well as ethylone (which was reported by DEA as most prevalent from seized compounds in 2015) more selectively bind to SERT compared to DAT.

8.5 Pharmacotherapeutic Avenues for Treating Abuse of Synthetic Cathinones

To date, no pharmacotherapies are available to treat dependence to psychostimulants, including cocaine, amphetamine, and synthetic cathinones. Only a few preclinical reports make effort to normalize behavioral effects following synthetic cathinone administration. Philogene-Khalid and colleagues (2017a) recently reported that *S*-mephedrone, which works at greater capacity than its racemate to augment 5-HT neurochemistry, acutely attenuates anxiety- and depression-like behavior otherwise emerging following chronic, binge-like cocaine, or MDPV. Similarly, *S*-mephedrone pretreatment suppressed cocaine place preference in an invertebrate assay (Vouga et al. 2015). The authors mention that numerous clinically available medications, including bupropion (Wellbutrin®) and amphetamine (Adderall), take advantage of stereochemistry and enantiomer-specific effects to unlock therapeutic benefit while minimizing, as best as possible, abuse potential.

Separately, numerous neuropeptide systems have been framed as reward-seeking modulators through extensive innervation within midbrain and striatal structures. The clinically available sleep aide suvorexant (Belsomra®), which works by blocking excitatory hypocretin/orexin (Hcrt/OX) receptors, was recently shown to significantly reduce positively valenced 50-kHz USVs emerging during pre- and post-drug time periods from MDPV self-administering rats (Simmons et al. 2017). Prior work from this team found that suvorexant suppressed cocaine-evoked DA efflux in ventral striatum relative to vehicle-pretreated control subjects, and that suvorexant dose-dependently reduced breakpoint for self-administered cocaine under PR access conditions (Gentile et al. 2018). While a significant effect of suvorexant on the number of MDPV infusions earned was not observed from this low-effort FR1 task, the authors speculate that Hcrt/OX receptor blockade tends to favorably reduce motivated responding for psychostimulants, and that suvorexant may exert greater and translationally valid therapeutic effect in high-effort assays including PR self-administration or when employing a behavioral-economics self-administration model.

Finally, pharmacologic restoration of extracellular glutamate content within central nervous system following synthetic cathinone use has produced therapeutic effects in preclinical models. Surges in glutamate transmission following drug-associated cue/context exposure are linked to psychostimulant relapse. Repeated administration of ceftriaxone, a widely available β-lactam antibiotic, enhances levels of the glutamate transporter GLT1, which itself is bound on astrocyte membranes and works to actively transport glutamate from extra- to intra-cellular compartments (Rothstein et al. 2005). Ceftriaxone was shown to normalize MDPV-induced locomotor sensitization as well as place preference to an MDPV-paired context (Gregg et al. 2016). Metabotropic glutamate receptors 2 and 3 (mGluR2/3) are presynaptically located and function to suppress glutamate release, thus acting in large part as autoreceptors. The mGluR2/3 endogenous agonist *N*-acetylaspartylglutamate (NAAG) is hydrolyzed by glutamate carboxypeptidase II (GCPII), and pharmacologic inhibition of GCPII by 2-(phosphonomethyl)-pentanedioic acid (2-PMPA) can enhance NAAG levels and in turn enhance mGluR2/3 activation via its endogenous ligand. Pretreatment with 2-PMPA was recently reported to significantly decrease MDPV place preference in rats (Hicks et al. 2017) although effects on locomotor activity following an acute MDPV injection were not observed. In sum, suppressing otherwise elevated levels of extracellular glutamate and/or glutamate transmission following chronic psychostimulant use has validated preclinical therapeutic efficacy in numerous models of drug-associated reward and reinforcement, and this treatment concept is being actively explored for management of substance use disorders in humans (e.g., see NCT02563769 on ClinicalTrials.gov).

Contributions and Acknowledgements SJS wrote and edited this chapter with input from EK, TAG, AM, and JWM under direction of SMR. The authors appreciate grant support from the National Institute on Drug Abuse (R01 DA039139 and R21 DA032718 to SMR; P30 DA013429 to Ellen M. Unterwald).

References

Aarde SM, Huang PK, Creehan KM et al (2013a) The novel recreational drug 3, 4-methylenedioxypyrovalerone (MDPV) is a potent psychomotor stimulant: self-administration and locomotor activity in rats. Neuropharmacology 71:130–140. https://doi.org/10.1016/j.neuropharm.2013.04.003

Aarde SM, Angrish D, Barlow DJ et al (2013b) Mephedrone (4-methylmethcathinone) supports intravenous self-administration in Sprague-Dawley and Wistar rats. Addict Biol 18(5):786–799. https://doi.org/10.1111/adb.12038

Ahmed SH, Koob GF (1998) Transition from moderate to excessive drug intake: change in hedonic set point. Science 282(5387):298–300

Antonowicz JL, Metzger AK, Ramanujam SL (2011) Paranoid psychosis induced by consumption of methylenedioxypyrovalerone: two cases. Gen Hosp Psychiatry 33(6):640.e5-6. https://doi.org/10.1016/j.genhosppsych.2011.04.010

Bajaj N, Mullen D, Wylie S (2010) Dependence and psychosis with 4-methylmethcathinone (mephedrone) use. BMJ Case Rep pii: bcr0220102780. https://doi.org/10.1136/bcr.02.2010.2780

Banjaw MY, Schmidt WJ (2005) Behavioural sensitisation following repeated intermittent oral administration of *Catha edulis* in rats. Behav Brain Res 156(2):181–189. https://doi.org/10.1016/j.bbr.2004.05.020

Banjaw MY, Miczek K, Schmidt WJ (2006) Repeated *Catha edulis* oral administration enhances the baseline aggressive behavior in isolated rats. J Neural Transm (Vienna) 113(5):543–556

Barker DJ, Simmons SJ, Servilio LC et al (2014) Ultrasonic vocalizations: evidence for an affective opponent process during cocaine self-administration. Psychopharmacology 231 (5):909–918. https://doi.org/10.1007/s00213-013-3309-0

Bauer CT, Banks ML, Blough BE, Negus SS (2013) Use of intracranial self-stimulation to evaluate abuse-related and abuse-limiting effects of monoamine releasers in rats. Br J Pharmacol 168(4):850–862. https://doi.org/10.1111/j.1476-5381.2012.02214.x

Baumann MH, Ayestas MA Jr, Partilla JS et al (2012) The designer methcathinone analogs, mephedrone and methylone, are substrates for monoamine transporters in brain tissue. Neuropsychopharmacology 37(5):1192–1203. https://doi.org/10.1038/npp.2011.304

Baumann MH, Partilla JS, Lehner KR et al (2013) Powerful cocaine-like actions of 3, 4-methylenedioxypyrovalerone (MDPV), a principal constituent of psychoactive 'bath salts' products. Neuropsychopharmacology 38(4):552–562. https://doi.org/10.1038/npp.2012.204

Berquist MD 2nd, Traxler HK, Mahler AM, Baker LE (2016) Sensitization to the locomotor stimulant effects of "bath salt" constituents, 4-methylmethcathinone (4-MMC) and 3, 4-methylenedioxypyrovalerone (MDPV), in male Sprague-Dawley rats. Drug Alcohol Depend 164:128–134. https://doi.org/10.1016/j.drugalcdep.2016.05.001

Berquist MD 2nd, Thompson NA, Baker LE (2017) Evaluation of training dose in male Sprague-Dawley rats trained to discriminate 4-methylmethcathinone. Psychopharmacology 234(2):3271–3278. https://doi.org/10.1007/s00213-017-4716-4

Bonano JS, Glennon RA, De Felice LJ et al (2014) Abuse-related and abuse-limiting effects of methcathinone and the synthetic "bath salts" cathinone analogs methylenedioxypyrovalerone (MDPV), methylone and mephedrone on intracranial self-stimulation in rats. Psychopharmacology 231(1):199–207. https://doi.org/10.1007/s00213-013-3223-5

Botanas CJ, Yoon SS, de la Peña JB et al (2017a) The abuse potential of two novel synthetic cathinones with modification on the alpha-carbon position, 2-cyclohexyl-2-(methylamino)-1-phenylethanone (MACHP) and 2-(methylamino)-1-phenyloctan-1-one (MAOP), and their effects on dopaminergic activity. Pharmacol Biochem Behav 153:160–167. https://doi.org/10.1016/j.pbb.2016.12.017

Botanas CJ, Yoon SS, de la Peña JB et al (2017b) A novel synthetic cathinone, 2-(methylamino)-1-(naphthalen-2-yl) propan-1-one (BMAPN), produced rewarding effects and altered striatal dopamine-related gene expression in mice. Behav Brain Res 317:494–501. https://doi.org/10.1016/j.bbr.2016.10.016

Botanas CJ, Yoon SS, de la Peña JB et al (2017c) The abuse potential of α-piperidinopropiophenone (PIPP) and α-piperidinopentiothiophenone (PIVT), two new synthetic cathinones with piperidine ring substituent. Biomol Ther (Seoul) 25(2):122–129. https://doi.org/10.4062/biomolther.2016.241

Burgdorf J, Knutson B, Panksepp J (2000) Anticipation of rewarding electrical brain stimulation evokes ultrasonic vocalization in rats. Behav Neurosci 14(2):320–327

Calcagnetti DJ, Schechter MD (1992) Increases in the locomotor activity of rats after intracerebral administration of cathinone. Brain Res Bull 29(6):843–846

Calcagnetti DJ, Schechter MD (1993) Place preference for the psychostimulant cathinone is blocked by pretreatment with a dopamine release inhibitor. Prog Neuropsychopharmacol Biol Psychiatry 17(4):637–649

Cheong JH, Choi MJ, Jang CG et al (2017) Behavioral evidence for the abuse potential of the novel synthetic cathinone alpha-pyrrolidinopentiothiophenone (PVT) in rodents. Psychopharmacology 234(5):857–867. https://doi.org/10.1007/s00213-017-4526-8

Creehan KM, Vandewater SA, Taffe MA (2015) Intravenous self-administration of mephedrone, methylone and MDMA in female rats. Neuropharmacology 92:90–97. https://doi.org/10.1016/j.neuropharm.2015.01.003

Dargan PI, Albert S, Wood DM (2010) Mephedrone use and associated adverse effects in school and college/university students before the UK legislation change. QJM 103(11):875–879. https://doi.org/10.1093/qjmed/hcq134

Derungs A, Schietzel S, Meyer MR et al (2011) Sympathomimetic toxicity in a case of analytically confirmed recreational use of naphyrone (naphthylpyrovalerone). Clin Toxicol (Phila) 49 (7):691–693. https://doi.org/10.3109/15563650.2011.592838

Dolengevich-Segal H, Rodríguez-Salgado B, Gómez-Arnau J, Sánchez-Mateos D (2016) Severe psychosis, drug dependence, and Hepatitis C related to slamming mephedrone. Case Rep Psychiatry. https://doi.org/10.1155/2016/8379562

Eshleman AJ, Wolfrum KM, Reed JF et al (2017) Structure-activity relationships of substituted cathinones, with transporter binding, uptake, and release. J Pharmacol Exp Ther 360(1):33–47. https://doi.org/10.1124/jpet.116.236349

Garrett G, Sweeney M (2010) The serotonin syndrome as a result of mephedrone toxicity. BMJ Case Rep pii: bcr0420102925. https://doi.org/10.1136/bcr.04.2010.2925

Gatch MB, Dolan SB, Forster MJ (2015a) Comparative behavioral pharmacology of three pyrrolidine-containing synthetic cathinone derivatives. J Pharmacol Exp Ther 354(2):103–110. https://doi.org/10.1124/jpet.115.223586

Gatch MB, Rutledge MA, Forster MJ (2015b) Discriminative and locomotor effects of five synthetic cathinones in rats and mice. Psychopharmacology 232(7):1197–1205. https://doi.org/10.1007/s00213-014-3755-3

Gentile TA, Simmons SJ, Barker DJ, Shaw JK, España RA, Muschamp JW (2018) Suvorexant, an orexin/hypocretin receptor antagonist, attenuates motivational and hedonic properties of cocaine. Addict Biol 23(1):247–255

Glennon RA, Martin BR, Dal Cason TA, Young R (1995) Methcathinone ("cat"): an enantiomeric potency comparison. Pharmacol Biochem Behav 50(4):601–606

Glennon RA, Yousif M, Naiman N, Kalix P (1987) Methcathinone: a new and potent amphetamine-like agent. Pharmacol Biochem Behav 26(3):547–551

Gosnell BA, Yracheta JM, Bell SM, Lane KE (1996) Intravenous self-administration of cathinone by rats. Behav Pharmacol 7(6):526–531

Gregg RA, Baumann MH, Partilla JS et al (2015) Stereochemistry of mephedrone neuropharmacology: enantiomer–specific behavioural and neurochemical effects in rats. Br J Pharmacol 172(3):883–894. https://doi.org/10.1111/bph.12951

Gregg RA, Hicks C, Nayak SU et al (2016) Synthetic cathinone MDPV downregulates glutamate transporter subtype I (GLT-1) and produces rewarding and locomotor-activating effects that are reduced by a GLT-1 activator. Neuropharmacology 108:111–119. https://doi.org/10.1016/j.neuropharm.2016.04.014

Gregg RA, Tallarida CS, Reitz A et al (2013) Mephedrone (4-methylmethcathinone), a principal constituent of psychoactive bath salts, produces behavioral sensitization in rats. Drug Alcohol Depend 133(2):746–750. https://doi.org/10.1016/j.drugalcdep.2013.06.014

Hadlock GC, Webb KM, McFadden LM et al (2011) 4-Methylmethcathinone (mephedrone): neuropharmacological effects of a designer stimulant of abuse. J Pharmacol Exp Ther 339 (2):530–536. https://doi.org/10.1124/jpet.111.184119

Hicks C, Gregg RA, Nayak SU et al (2017) Glutamate carboxypeptidase II (GCPII) inhibitor 2-PMPA reduces rewarding effects of the synthetic cathinone MDPV in rats: a role for N-acetylaspartylglutamate (NAAG). Psychopharmacology 234(11):1671–1681. https://doi.org/10.1007/s00213-017-4568-y

Huskinson SL, Naylor JE, Townsend EA et al (2017) Self-administration and behavioral economics of second-generation synthetic cathinones in male rats. Psychopharmacology 234 (4):589–598. https://doi.org/10.1007/s00213-016-4492-6

Hutsell BA, Baumann MH, Partilla JS et al (2016) Abuse-related neurochemical and behavioral effects of cathinone and 4-methylcathinone stereoisomers in rats. Eur Neuropsychopharmacol 26(2):288–297. https://doi.org/10.1016/j.euroneuro.2015.12.010

Iversen L, Gibbons S, Treble R et al (2013) Neurochemical profiles of some novel psychoactive substances. Eur J Pharmacol 700(1–3):147–151. https://doi.org/10.1016/j.ejphar.2012.12.006

Johnson PS, Johnson MW (2014) Investigation of "bath salts" use patterns within an online sample of users in the United States. J Psychoactive Drugs 46(5):369–378. https://doi.org/10.1080/02791072.2014.962717

Kaizaki A, Tanaka S, Numazawa S (2014) New recreational drug 1-phenyl-2-(1-pyrrolidinyl)-1-pentanone (alpha-PVP) activates central nervous system via dopaminergic neuron. J Toxicol Sci 39(1):1–6

Kalix P (1980) A constituent of khat leaves with amphetamine-like releasing properties. Eur J Pharmacol 68(2):213–215

Kalix P (1981) Cathinone, an alkaloid from khat leaves with an amphetamine-like releasing effect. Psychopharmacology 74(3):269–270

Karlsson L, Andersson M, Kronstrand R, Kugelberg FC (2014) Mephedrone, methylone and 3,4-methylenedioxypyrovalerone (MDPV) induce conditioned place preference in mice. Basic Clin Pharmacol Toxicol 115(5):411–416. https://doi.org/10.1111/bcpt.12253

Kehr J, Ichinose F, Yoshitake S et al (2011) Mephedrone, compared with MDMA (ecstasy) and amphetamine, rapidly increases both dopamine and 5-HT levels in nucleus accumbens of awake rats. Br J Pharmacol 164(8):1949–1958. https://doi.org/10.1111/j.1476-5381.2011.01499.x

Knoll J (1979) Studies on the central effects of (-) cathinone. NIDA Res Monogr 27:322–323

Knutson B, Burgdorf J, Panksepp J (2002) Ultrasonic vocalizations as indices of affective states in rats. Psychol Bull 128(6):961–977

Kolanos R, Solis E Jr, Sakloth F et al (2013) "Deconstruction" of the abused synthetic cathinone methylenedioxypyrovalerone (MDPV) and an examination of effects at the human dopamine transporter. ACS Chem Neurosci 4(12):1524–1529. https://doi.org/10.1021/cn4001236

Kolanos R, Partilla JS, Baumann MH et al (2015) Stereoselective actions of methylenedioxypyrovalerone (MDPV) to inhibit dopamine and norepinephrine transporters and facilitate intracranial self-stimulation in rats. ACS Chem Neurosci 6(5):771–777. https://doi.org/10.1021/acschemneuro.5b00006

López-Arnau R, Martínez-Clemente J, Pubill D et al (2012) Comparative neuropharmacology of three psychostimulant cathinone derivatives: butylone, mephedrone and methylone. Br J Pharmacol 167(2):407–420. https://doi.org/10.1111/j.1476-5381.2012.01998.x

Ma ST, Maier EY, Ahrens AM et al (2010) Repeated intravenous cocaine experience: development and escalation of pre-drug anticipatory 50-kHz ultrasonic vocalizations in rats. Behav Brain Res 212(1):109–114. https://doi.org/10.1016/j.bbr.2010.04.001

Marusich JA, Grant KR, Blough BE, Wiley JL (2012) Effects of synthetic cathinones contained in "bath salts" on motor behavior and a functional observational battery in mice. Neurotoxicology 33(5):1305–1313. https://doi.org/10.1016/j.neuro.2012.08.003

Marusich JA, Lefever TW, Blough BE et al (2016) Pharmacological effects of methamphetamine and alpha-PVP vapor and injection. Neurotoxicology 55:83–91. https://doi.org/10.1016/j.neuro.2016.05.015

Mohammed A, Engidawork E (2011) Reproductive parameters are differentially altered following subchronic administration of Catha edulis F. (Khat) extract and cathinone in male rats. J Ethnopharmacol 134(3):977–983. https://doi.org/10.1016/j.jep.2011.02.006

Motbey CP, Clemens KJ, Apetz N et al (2013) High levels of intravenous mephedrone (4-methylmethcathinone) self-administration in rats: neural consequences and comparison with methamphetamine. J Psychopharmacol 27(9):823–836. https://doi.org/10.1177/0269881113490325

Naylor JE, Freeman KB, Blough BE et al (2015) Discriminative-stimulus effects of second generation synthetic cathinones in methamphetamine-trained rats. Drug Alcohol Depend 149:280–284. https://doi.org/10.1016/j.drugalcdep.2015.02.002

Nelson KH, Hempel BJ, Clasen MM et al (2017) Conditioned taste avoidance, conditioned place preference and hyperthermia induced by the second generation 'bath salt' α-pyrrolidinopentiophenone (α-PVP). Pharmacol Biochem Behav 156:48–55. https://doi.org/10.1016/j.pbb.2017.04.003

Nguyen JD, Aarde SM, Cole M et al (2016) Locomotor stimulant and rewarding effects of inhaling methamphetamine, MDPV, and mephedrone via electronic cigarette-type technology. Neuropsychopharmacology 41(11):2759–2771. https://doi.org/10.1038/npp.2016.88

Pehek EA, Schechter MD, Yamamoto BK (1990) Effects of cathinone and amphetamine on the neurochemistry of dopamine in vivo. Neuropharmacology 29(12):1171–1176

Philogene-Khalid HL, Hicks C, Reitz AB et al (2017a) Synthetic cathinones and stereochemistry: S enantiomer of mephedrone reduces anxiety-and depressant-like effects in cocaine-or MDPV-abstinent rats. Drug Alcohol Depend 178:119–125. https://doi.org/10.1016/j.drugalcdep.2017.04.024

Philogene-Khalid HL, Simmons SJ, Nayak S et al (2017b) Stereoselective differences between the reinforcing and motivational effects of cathinone-derived 4-methylmethcathinone (mephedrone) in self-administering rats. ACS Chem Neurosci. https://doi.org/10.1021/acschemneuro.7b00212

Roberts L, Ford L, Patel N et al (2017) 11 analytically confirmed cases of mexedrone use among polydrug users. Clin Toxicol (Phila) 55(3):181–186. https://doi.org/10.1080/15563650.2016.1271424

Robinson JE, Agoglia AE, Fish EW et al (2012) Mephedrone (4-methylmethcathinone) and intracranial self-stimulation in C57BL/6 J mice: comparison to cocaine. Behav Brain Res 234(1):76–81. https://doi.org/10.1016/j.bbr.2012.06.012

Saha K, Partilla JS, Lehner KR et al (2015) 'Second-generation' mephedrone analogs, 4-MEC and 4-MePPP, differentially affect monoamine transporter function. Neuropsychopharmacology 40(6):1321–1331. https://doi.org/10.1038/npp.2014.325

Schechter MD (1991) Effect of learned behavior upon conditioned place preference to cathinone. Pharmacol Biochem Beh 38(1):7–11

Schindler CW, Thorndike EB, Goldberg SR et al (2016) Reinforcing and neurochemical effects of the "bath salts" constituents 3, 4-methylenedioxypyrovalerone (MDPV) and 3, 4-methylenedioxy-N-methylcathinone (methylone) in male rats. Psychopharmacology 233(10):1981–1990. https://doi.org/10.1007/s00213-015-4057-0

Schuster CR, Johanson CE (1979) Behavioural studies of cathinone in monkeys and rats. NIDA Res Monogr 27:324–325

Shortall SE, Macerola AE, Swaby RT et al (2013) Behavioural and neurochemical comparison of chronic intermittent cathinone, mephedrone and MDMA administration to the rat. Eur Neuropsychopharmacol 23(9):1085–1095. https://doi.org/10.1016/j.euroneuro.2012.09.005

Simmler LD, Buser TA, Donzelli M et al (2013) Pharmacological characterization of designer cathinones *in vitro*. Br J Pharmacol 168(2):458–470. https://doi.org/10.1111/j.1476-5381.2012.02145.x

Simmler LD, Rickli A, Hoener MC, Liechti ME (2014) Monoamine transporter and receptor interaction profiles of a new series of designer cathinones. Neuropharmacology 79:152–160. https://doi.org/10.1016/j.neuropharm.2013.11.008

Simmons SJ, Gregg RA, Tran FH et al (2016) Comparing rewarding and reinforcing properties between 'bath salt'3, 4-methylenedioxypyrovalerone (MDPV) and cocaine using ultrasonic vocalizations in rats. Addict Biol. https://doi.org/10.1111/adb.12479

Simmons SJ, Martorana R, Philogene-Khalid H et al (2017) Role of hypocretin/orexin receptor blockade on drug-taking and ultrasonic vocalizations (USVs) associated with low-effort self-administration of cathinone-derived 3, 4-methylenedioxypyrovalerone (MDPV) in rats. Psychopharmacology 234(21):3207–3215. https://doi.org/10.1007/s00213-017-4709-3

Spiller HA, Ryan ML, Weston RG, Jansen J (2011) Clinical experience with and analytical confirmation of "bath salts" and "legal highs"(synthetic cathinones) in the United States. Clin Toxicol (Phila) 49(6):499–505. https://doi.org/10.3109/15563650.2011.590812

Taha SA, Ageel AM, Islam MW, Ginawi OT (1995) Effect of (-)-cathinone, a psychoactive alkaloid from khat (*Catha edulis* Forsk.) and caffeine on sexual behaviour in rats. Pharmacol Res 31(5):299–303

Tzschentke TM (1998) Measuring reward with the conditioned place preference paradigm: a comprehensive review of drug effects, recent progress and new issues. Prog Neurobiol 56(6):613–672

Van Hout MC, Bingham T (2012) "A costly turn on": patterns of use and perceived consequences of mephedrone based head shop products amongst Irish injectors. Int J Drug Policy 23(3):188–197. https://doi.org/10.1016/j.drugpo.2012.01.008

Vouga A, Gregg RA, Haidery M et al (2015) Stereochemistry and neuropharmacology of a 'bath salt' cathinone: S-enantiomer of mephedrone reduces cocaine-induced reward and withdrawal in invertebrates. Neuropharmacology 91:109−116

Watterson LR, Hood L, Sewalia, K, Tomek SE et al (2012a) The reinforcing and rewarding effects of methylone, a synthetic cathinone commonly found in "bath salts". J Addict Res Ther 9. pii: 002. https://doi.org/10.4172/2155-6105.s9-002

Watterson LR, Kufahl PR, Nemirovsky NE et al (2012b) Potent rewarding and reinforcing effects of the synthetic cathinone 3, 4-methylenedioxypyrovalerone (MDPV). Addict Biol 19(2):165–174. https://doi.org/10.1111/j.1369-1600.2012.00474.x

Watterson LR, Burrows BT, Hernandez RD, Moore KN, Grabenauer M, Marusich JA, Olive MF (2015) Effects of α-Pyrrolidinopentiophenone and 4-Methyl-N-Ethylcathinone, two synthetic cathinones commonly found in second-generation "bath salts," on intracranial self-stimulation thresholds in rats. Int J Neuropsychopharmacol 18(1). https://doi.org/10.1093/ijnp/pyu014

Woolverton WL, Johanson CE (1984) Preference in rhesus monkeys given a choice between cocaine and d, l-cathinone. J Exp Anal Behav 41(1):35–43

Chapter 9
Synthetic Cathinones—Prevalence and Motivations for Use

John M. Corkery, Amira Guirguis, Duccio G. Papanti, Laura Orsolini and Fabrizio Schifano

Abstract This chapter considers the prevalence of and motivations for use of synthetic cathinones. As part of the scene-setting, the availability, legal status, numbers of cathinones, number and quantities confiscated are reviewed. This leads to the first substantive section of the chapter—an epidemiological investigation of the nature and extent of what is known about the use of these molecules. The second major section is more qualitative in its approach to understanding motivations for the use of any drug, Novel Psychoactive Substances (NPS), and then synthetic cathinones. An examination is conducted of how cathinones may be compared to other stimulants and why particular cathinones may be preferred to others. The converse situation is then examined, what might be the motivations and reasons for ceasing to take cathinones and why this may not be a rational decision. A brief examination of the consequences of ceasing versus continued use is presented. As it is very likely that further synthetic cathinones will continue to emerge, it is important to gain a much fuller insight into what motivates or causes individuals to use or cease using these molecules, so that communities and societies can respond in appropriate ways to the varying challenges that face them and their citizens.

Keywords Synthetic cathinones · Epidemiology · Prevalence · Reasons Motivations

9.1 Introduction and Background

Despite the continuing increase in the emergence of Novel Psychoactive Substances (NPS), a core group of over 80 NPS has become established as attractive recreational drugs, supplementing traditional drugs of abuse and becoming part of the

J. M. Corkery (✉) · A. Guirguis · D. G. Papanti · L. Orsolini · F. Schifano
Psychopharmacology, Drug Misuse and Novel Psychoactive Substances Research Unit, School of Life and Medical Sciences, University of Hertfordshire, Hatfield, UK
e-mail: j.corkery@herts.ac.uk

J. B. Zawilska (ed.), *Synthetic Cathinones*, Current Topics in Neurotoxicity 12,
https://doi.org/10.1007/978-3-319-78707-7_9

repertoire of substances available for consumption (Moore et al. 2013; UNODC 2017a). This group has been reported every year from 2009 to 2016, including mephedrone and derivatives. In Central Asia, this group accounts for 42% of all new NPS notifications in 2013–2016 compared to 16% of NPS notified from the Near and Middle East since 2008 (UNODC 2017b). It is important, therefore, to understand the prevalence and motivations for use of the largest NPS groups, including synthetic cathinones (EMCDDA 2016a, 2017; UNODC 2016a), in order to appropriately understand the nature and extent of the problem. This will facilitate the design of better-informed data collection models, as well as the delivery of appropriately tailored harm-reduction techniques and initiatives.

Estimating demand for and supply of psychoactive substances is problematic even for traditional drugs of abuse, let alone for NPS (Corkery et al. in press). For example, the number of NPS confiscations or the amounts of NPS seized by law enforcement agencies only reflect operational priorities by such organisations. Furthermore, only very inaccurate estimates can be made as to what proportion of the total supply of specific substances is accounted for by such law enforcement agency interdictions. On the demand side, very little is known about the nature of NPS consumption, especially with regard to typical doses taken to achieve particular effects, the accuracy of dose measurement, the frequency of dosing, need to re-dose, tolerance, duration of effects, length of half-life, metabolism, toxic and fatal levels, etc.

However, international agencies such as the United Nations Office for Drugs and Crime (UNODC) and the European Monitoring Centre for Drugs and Drug Addiction (EMCDDA) have work programmes aiming to enhance the development of more timely and robust techniques to not only monitor the current situation with regard to the NPS phenomenon but also predict future evolutions. Indeed, new legislation came into force across the EU on 22 November 2017, and will become applicable 12 months later; it is aimed at speeding up the notification and assessment processes (OJEU 2017). In the meantime, we have to deal with the limitations and vagaries of the established monitoring systems and the challenges which face them (Palamar et al. 2015; UNODC 2017a). Whilst patterns can be observed in the number and type of NPS being reported to such international agencies, there are limitations in terms of the timeliness of detection, accurate identification and submission of reports by law enforcement, forensic and toxicological services. Furthermore, whilst we may have a good estimate through these indicators of the number of NPS in circulation, the actual numbers that have been created and held ready for sale are likely to be higher.

It is very challenging to accurately estimate the prevalence of specific NPS classes because most of the surveys conducted focus either on NPS as a whole or on specific NPS, although some progress has been made in recent years in improving survey instruments and tools (UNODC 2017a). Other technical issues include self-reporting surveys, such as the Global Drug Survey (Winstock et al. 2017), may not be representative of all users; there may be differences between respondents who have Internet access and chose to participate compared to those who lack such access or desire to take part (Ashrafioun et al. 2016). General household surveys which will not capture people living alone, in institutional settings (e.g. hospitals, penal establishments, student accommodation, and sheltered housing) or those who are 'homeless' or 'roofless' resorting to 'sofa-surfing'—who often comprise individuals most at risk of using and/or dependent on psychoactive substances, including synthetic cathinones. It often takes a relatively long period to get specific questions on special topics included in regular surveys, such as the Crime Survey for England and Wales (Broadfield 2017), so that by the time the survey is in the field the substances being asked about may have disappeared from the market. Palamar et al. (2017) argue that 'gate' questions, which adopt a 'skip-logic' approach in surveys such that 'only a "yes" response to the use of specific drug class is followed by more extensive queries of drug use in that drug class', may impact the accuracy of collected data via self-reporting. Therefore, surveys on NPS use should explicitly include the NPS classes, the specific NPS within each class as well as the street names.

NPS users often have little knowledge about what compounds they are actually consuming; they may only be aware of a brand name or a 'street' name, and almost certainly will not have had it analysed. That said, often products purchased as 'research chemicals' are indeed what they purport to be (Guirguis et al. 2017).

In this chapter, we focus on the second largest group of NPS—synthetic cathinones. As noted elsewhere in this book, these substances are synthetic derivatives of the natural amphetamine-like substances (cathine and cathinone) present in the leaves and stalks of the khat plant (*Catha edulis*), which produce mild stimulant and other psychoactive effects (Corkery 2016; see also Chap. 3).

9.2 Availability of Synthetic Cathinones

Before taking a closer look at the information available about the extent of use of synthetic cathinones, we will briefly review the situation regarding their availability. In Europe, by the end of 2016, some 629 NPS of all classes were being monitored by the EMCDDA, of which 66 were reported for the first time in 2016 (EMCDDA-Europol 2017). By the end of 2016, 739 substances had been reported by 106 countries and territories through the global Synthetics Monitoring: Analysis, Reporting and Trends (SMART) programme to the UNODC (2017a); the number of NPS reported increased from 126 to 450 between 2009 and 2014.

9.2.1 Legal Status of Cathinones

Khat is regarded as 'a particular NPS of concern' by the UNODC (2014). Although the plant itself is not under international control, at present, two of the principal psychoactive substances that are present in it, cathine and cathinone are. This is because in the early 1980s all amphetamine-like substances were placed group-wise under international control (ECCD 1985). Cathinone was included in Schedule I of the UN Convention on Psychotropic Substances in 1988, and cathine was then included in Schedule III of this Convention. Cathinone and cathine are controlled substances under Class C of the UK Misuse of Drugs Act 1971 by virtue of Schedules 1 and 3, respectively, of the Misuse of Drugs (Amendment) Regulations Act 1986.

A number of persistent NPS have been brought under international control, among them several cathinones: α-pyrrolidinovalerophenone (α-PVP) was placed under international control in 2016; 3,4-methylenedioxypyrovalerone (MDPV), mephedrone and methylone in 2015, all under Schedule II of the 1971 Convention on Psychotropic Substances (UNODC 2017a). In March 2017, the Commission on Narcotic Drugs decided to add ethylone, pentedrone and 4-methylethcathinone (4-MEC) to the same Schedule. In the USA, MDPV, mephedrone and methylone come within the compass of Schedule I of the Controlled Substances Act 1970.

At the EU level, mephedrone was submitted to control measures in EU Member States by the European Council's decision of 2 December 2010 (2010/759/EU). Some cathinone derivatives are caught by drug control or equivalent legislation, for example: mephedrone (Belgium, Crotia, Denmark, Estonia, France, Germany, Ireland, Italy, Lithuania, Norway, Romania and Sweden); methylone (Denmark, Ireland, Romania and Sweden); butylone (Denmark, Ireland, Norway, Romania and Sweden); MDPV (Denmark, Ireland, Finland and Sweden); and flephedrone (Denmark, Ireland and Romania). Medicines' legislation is used in Finland and the Netherlands to control mephedrone. In the UK context, several generations/classes of synthetic cathinones and derivatives have been dealt with by means of generic definitions under the 1971 Act as Class B drugs and under Schedule 1 of the Misuse of Drugs Regulations (2003, as amended).

Some synthetic cathinones are licensed for medicinal use. For example, bupropion (Zyban, Wellbutrin), primarily used as an antidepressant and smoking cessation aid, has been used recreationally (Vento et al. 2013). Pyrovalerone, developed in the late 1960s and structurally related to MDPV, has been used in the clinical treatment of chronic fatigue and lethargy due to its stimulant effects (Gardos and Cole 1971), as well as an appetite suppressant and anorectic in countries such as France. However, it is not often prescribed these days because of issues with abuse and dependence (Deniker et al. 1975). The molecule is covered by Schedule IV of the 1971 UN Convention; it was controlled as a Class C drug under the Misuse of Drugs Act 1971 prior to the inclusion of other cathinones in April 2010.

In the past, synthetic cathinones could be purchased through a range of legal retail outlets, including convenience stores, gas/petrol stations, 'head'/'smart' shops, 'adult' stores, etc. (Meyers et al. 2015). Over the last decade, the Internet has become a major nexus for information on, and a leading marketplace for the purchase of NPS in general, but also for this group of psychoactive compounds. In fact, the anonymity of the Internet has often been proposed as a reason for the growth in the use of synthetic cathinones. However, the imposition of so-called 'blanket bans' by national governments, e.g. the Republic of Ireland and the UK, has not only led to the closure of retail outlets such as 'head'/'smart' shops and an open source of NPS, but has driven the market for NPS to the clandestine street market alongside traditional drugs such as amphetamine, cannabis, cocaine and heroin (Shapiro and Daly 2017), as well as to the deeper recesses of the Internet. Wadsworth et al. (2017) report a possible shift of cathinone sales in the UK from the surface web to the deep web, following the coming into effect of the Psychoactive Substances Act 2016 on 26 May 2016.

9.2.2 Number of Synthetic Cathinones

At a global level, the UNODC reported that the number of synthetic cathinones identified/reported by member countries and territories rose from 7 in 2008 to 68 by the end of 2014, with about 55 in 2015 (provisional data) (UNODC 2016b). Over the period 2005–2016, the cumulative number of such compounds reported to the EMCDDA was 118, of which 14 were reported for the first time in 2016 (EMCDDA-Europol 2017); by the end of 2017 this figure stood at 130.

9.2.3 Number and Quantities of Cathinones Confiscated

The fact that most NPS remain outside international drug control and cannot, therefore, be seized and reported to the UNODC as part of the regular data collection mechanisms, such as the Annual Report Questionnaire, means that is difficult to get an accurate picture of their movement and supply. To complicate matters further, confiscations of NPS are commonly only notified using a product name or at the substance group level, which prevents data from being linked to a specific substance.

Very few seizures of synthetic NPS were reported in 2010, but the UNODC (2017a) reports that in recent years 23–54 tons have been seized annually. Whilst global seizure amounts have been dominated since 2011 by synthetic cannabinoids, the amount of synthetic cathinones confiscated has risen greatly since they were first reported in 2010. In 2010 and 2011 less than 200 kg were seized, in 2012 and 2013 the quantities doubled to about 400 kg, but seizures trebled between 2013 and 2014, reaching a level of 1.3 tons (UNODC 2016a). In 2014, most synthetic cathinones were seized in Eastern Europe (692 kg in the Russian Federation), in Western and Central Europe (312 kg in England and Wales) and in East and Southeast Asia (226 kg in Hong Kong, China).

The most recent US figures indicate that the number of cathinone 'exhibits' submitted to the National Forensic Laboratory Information System (NFLIS) increased from 15,523 in 2014 to 19,490 in 2015, ethylone accounting for 47% of synthetic cathinone exhibits (DEA 2016). During 2016 most cathinone identifications conducted by the U.S. Department of State Drug Enforcement Agency (DEA) Special Testing and Research Laboratory were for dibutylone (22%), followed by ethylone, *N*-ethylpentylone and α-PVP; these substances still appear to be most commonly identified in the first three quarters of 2017, although there is some variation from quarter to quarter (DEA 2016, 2017a, b, c).

The EMCDDA (2017) reports that for the latest annual reporting year (2015) some 80,000 seizures of NPS were made through the EU Early Warning System (EWS). Synthetic cathinones accounted for the largest class of NPS, some one-third or 25,000 + seizures, an increase of more than 17,000 in 2014. The quantity of synthetic cathinones seized in 2015 was just over 1.8 tonnes (1.05 tonnes in 2014). According to the EU Early Warning System (EWS), the five most commonly seized cathinones in 2015 were α-pyrrolidinopentiophenone (pyrrolidinovalerophenone; α-PVP), 3-methylmethcathinone (3-MMC), ethylone, 4-chloromethcathinone (4-CMC) and pentedrone. Where information on country of origin is available, China accounted for more than 1.2 tonnes of the synthetic cathinones seized in 2015. Of these, 42% were 2-methylmethcathinone (2-MMC) (156 kg) and 3-MMC (616 kg). These compounds are chemically related to mephedrone (4-MMC), but are not internationally controlled. The EMCDDA (2017) considers it likely that supplies of these two molecules are being passed off as mephedrone.

The number of mephedrone seizures made by law enforcement agencies in England and Wales rose from 2002 in 2010/2011 (the first full year following control) to a peak of 3850 in 2012/2013 since when the number has steadily fallen to 1031 in 2015/2016. Over the same period, the amount of mephedrone confiscated has followed a similar trajectory: from 231 kg in 2010/2011 to 716 kg in 2012/2013, falling to 173 kg in 2015/2016 (personal communication to first author by J. Hargreaves, Home Office Crime and Policing Analysis Unit, 24 November 2016). In Scotland, the police made around 200 seizures of mephedrone each year between 2010/2011 and 2012/2013 but this has since fallen to only 11 in 2015/2016. The amount seized rose from less than 6 kg in 2010/2011 to 9–10 kg in the following 2 years, but reached a peak of about 67 kg in 2013/2014 before falling to less than 1 kg in 2015/2016 (personal communication to first author by H. Mackenzie, Justice Analytical Services, Scottish Government, 3 July 2017). In Northern Ireland, the number of seizures of mephedrone powder fell steadily from 286 in 2010/2011 to 46 in 2015/2016. The amount of mephedrone seized in 2010/2011 was 20.9 kg but fell to between 2.3 and 3.9 kg in 2011/2012 to 2013/2014 before falling to 0.4 kg in 2014/2015 and 0.1 kg in 2015/2016 (personal communication to the first author by PSNI Statistics Branch, 2 December 2016).

9.3 Prevalence of Use

Although mephedrone was first synthesised 90 years ago (Saem de Burnaga Sanchez 1929), the first use of synthetic cathinones for recreational purposes seems to have been in the mid-2000s. The first wave or generation of synthetic cathinones to come to the attention of the general public, law enforcement and healthcare professionals was methylone in 2005 and mephedrone in 2007. This was at a time when the availability and purity of stimulants such as amphetamines, cocaine and ecstasy (MDMA) were falling and prices were going up, especially in the UK. Although synthesised in 1969 (http://catbull.com/alamut/Bibliothek/Boehringer_MDPV_Patent.htm), MDPV first emerged onto the recreational market in 2009; the number of pyrovalerone derivatives has increased greatly in recent years (Beck et al. 2017). The second generation of synthetic cathinones, such as naphyrone and 4-methyl-*N*-ethcathinone (4-MEC), emerged as replacements for mephedrone, methylone and MDPV when these became controlled in 2010. By this time, the so-called third-generation synthetic cathinones also began to emerge: 3,4-dimethylmethcathinone (3,4-DMMC), then pentedrone (α-methylaminovalerophenone) followed by α-PVP (Valente et al. 2014).

Information on the availability of synthetic cathinones has provided some context against which we can examine the extent of their use. However, there are limitations in respect of the sources we can use to try and achieve this aim, especially with regard to trend data.

9.3.1 International Sources

At the global level, the UNODC has not published any prevalence data for synthetic cathinones as a drug class or for individual molecules. However, some national data are available for a few countries.

In the USA little information has been collected on adult cathinone use (Stogner and Miller 2013). However, the Monitoring the Future survey has been monitoring the drug use of college students and adults aged 19–55 years since 1975. Rather than looking at synthetic cathinones, they monitor 'bath salts'; however, questions were first introduced only in 2012. This was the first national survey in the USA to provide information on bath salts. Whilst the last year use rate was reported in 2012 at 0.8, 0.6 and 1.3% in grades 8, 10 and 12, respectively, by 2016 the levels had fallen to less than 1% in all three age groups (Schulenberg et al. 2017). Whilst 0.8% of high school seniors reported using bath salts in the previous year, the level amongst young adults (19–30 years) was typically 0.3%. In 2012, the annual prevalence rate for males aged 19–22 years was 3.0% compared to 0.5% amongst females of the same age. However, this difference disappeared in 2013, at a time when there was a large rise in the perceived risk of harm from using bath salts; such a shift was also seen in older groups of young adults, as well as 12th graders (18 years). The fall in the use of bath salts and a rise in their perceived risk of causing harm echoes US National Poison Control Centers' data suggestive of a peak occurring in 2012 of their use, with the subsequent decline likely to be due to the wide coverage of adverse effects in the media. There is currently little variation in last year use of baths salts across the different regions of the USA. In 2012, use of bath salts appears to have been higher in small towns and farm/country areas, particularly amongst young people, but now rates are very low (less than 1%) across all levels of population density. Use appears to be higher amongst non-college respondents (0.5%) than college students (<0.05%) in 2016. A study by Stogner and Miller (2013) of 2349 students at a Southeastern university in early 2012 found that only 1.07% reported lifetime use of bath salts or MDPV; use was higher amongst men (1.7%) than women (0.5%). In terms of ethnicity, the highest rates were for Hispanics (4.7%), followed by Native Americans (4.3%), Whites (0.9%) and Blacks (0.7%). Use was more common amongst athletes (4.0%) compared to non-athletes (0.9%).

In Australia, The Ecstasy and Related Drugs Reporting System (EDRS) has been running since 2003. EDRS is a national monitoring system intended to identify emerging trends of local and national interest in the markets for such drugs. EDRS looks at the price, purity, availability and patterns of use of ecstasy, methamphetamine, cocaine, ketamine, γ-hydroxybutyric acid (GHB), methylenedioxyamphetamine (MDA) and LSD. However, it also captures some information on mephedrone. The 2010 sweep of regular ecstasy users, the first to record a synthetic cathinone, found that 21% reported lifetime use of the molecule and 17% had used it in the 6 months prior to interview (Matthews and Bruno 2010). This was against a background of increased difficulty in obtaining ecstasy and lower purity levels. Recent use of synthetic cathinones fell from a peak of 18.5% in 2010 to 7.7% in 2015 (Sutherland et al. 2016). Recent use of mephedrone fell from 16% in 2010 to 13% in 2011 to 5% in 2012 and 6% in 2013, whilst use of MDPV increased from <1 to 1% between 2010 and 2011 before increasing to 2.5% in 2012 before falling to back to 1% in 2013; methylone use was 5% in 2011 and 2012, but fell to 3% in 2013 (Sindicich and Burns 2011, 2013). Based on multivariable logistic regression analysis of the 2010–2015 data, recent cathinone use was found to be associated with daily tobacco and poly-drug use. The combined intake of cathinones and cocaine was not statistically significant in the study by Sindicich and Burns (2011). Among poly-NPS users, in decreasing order of prevalence, recent cathinone use was associated with the intake of phenethylamines, tryptamines, synthetic cannabinoid receptor agonists (SCRAs), plants and extracts, arylcyclohexamines, piperazines and aminoindanes (Sutherland et al. 2016).

The EMCDDA has not yet collated any prevalence data for NPS; although whatever information that exists on specific molecules is presented in their Risk Assessment and EMCDDA-Europol Joint Report publication series (available at: http://www.emcdda.europa.eu/activities/action-on-new-drugs); those currently available on synthetic cathinones only relate to mephedrone, MDPV, and α-PVP. It was estimated in 2012–2013 that last year mephedrone use amongst European adults aged 16–59 years reached 0.5% (cited in Karila et al. 2016). Last year use of synthetic cathinones amongst young people aged 15–24 in Finland in 2014 was estimated to be about 0.2% (cited in Karila et al. 2017). In France MDPV and 4-MEC were in the top 10 NPS sold online in 2010–2011 (cited in Karila et al. 2017). In 2012, methylone was the 11th most popular hallucinogen in the USA (DEA 2012). In the Republic of Ireland, 3% of clients at a methadone maintenance clinic tested positive for methylone (McNamara et al. 2010). Last year use of injecting methylone was reported by 7% of male and 1% of female Irish prisoners in 2011 (Drummond et al. 2014).

In Europe, the Eurobarometer surveys on the attitudes of youth/young peoples' towards drugs only looked at 'new substances that imitate the effects of illicit drugs' (http://ec.europa.eu/justice/anti-drugs/new-drugs/index_en.htm). Also at the European level, there are no international surveys, such as European School Survey Project on Alcohol and Other Drugs (http://espad.org/) and the Health Behaviour in School-aged Children (http://www.hbsc.org/) capturing information on the use of school-age children on either specific synthetic cathinones or these molecules as a drug class.

At the UK level, the Smoking, Drinking, Drug Use survey carried out on behalf of the Department of Health (England) has tackled this issue. The most recent sweep found that last year use of mephedrone fell from 0.7% in 2012 to 0.5% in 2013 and 2014, and then to 0.4% in 2016 for pupils aged 11–15 years; the rate starting off higher for boys (0.8%) compared to that for girls (0.6%) in 2012, but both having the same rate (0.5%) in 2014, before falling to 0.4% for boys and 0.1% for girls in 2016 (NHS Digital 2017). In 2016, last year prevalence rates increased with age, as might be expected, for males from 0.0% for those aged 11 to 1.0% for 15 year-olds; for girls, it rose from 0.0% aged 11 and 12 years to 0.5% at 15 years. Offers of mephedrone in the last 12 months also increased with age, with 1% of boys aged 12 first experiencing offers compared to 4% of 15-year-old boys; for girls the age at when they were first offered mephedrone was 13 (1%), increasing to 3% by the age of 15 years. Whilst one-quarter (24%) of pupils aged 11 had ever tried mephedrone, by the age of 15 years this proportion had increased to 44% in 2016. In Scotland, a survey of students in high schools and tertiary educational institutions conducted a few years earlier found that lifetime use of mephedrone was 20.3%, 4.4% took it daily and 7.6% reported experiencing symptoms of dependence or addiction (Dargan et al. 2010).

9.3.2 *Mixmag/Global Drug Survey*

One of the longest running surveys of young people and their recreational drug-use patterns is the UK Mixmag drug survey associated with the dance music scene, which in recent years has expanded to become the Global Drug Survey (GDS). Though international in reach, most respondents are from the USA or the UK. The most obvious limitation is that participants are self-selected, thus subject to response bias (i.e. those interested in a subject are more likely to respond than those who are not) and may not represent all those young people who use drugs recreationally. The GDS uses non-purposive, non-random, opportunistic sampling methods.

As with other surveys, mephedrone was the first synthetic cathinone to be reported by readers of the Mixmag clubbing magazine in its 2010 survey (conducted in 2009), being the fourth most popular drug; lifetime use was reported at

41.7% and last month use at 33.6%. It was the sixth most common substance used in the month prior to the survey, only bettered by tobacco, alcohol, cannabis, cocaine and MDMA. Mephedrone users tended to be younger and male compared to other respondents. Just over half (54.6%) of individuals who had also used cocaine reported that a mephedrone high was of better quality; those using intranasally reported that mephedrone was more addictive and risky than cocaine (Winstock et al. 2011a). Lifetime use of mephedrone peaked in 2011 at 61%, as did last year use at 51%; these fell to 42.7 and 19.5%, respectively, in 2012, by which time last month use had fallen to 13%. However, last year use by regular UK clubbers was 30% (Winstock 2012), mephedrone use continued to decline in 2013: lifetime use was 36.1% and last year use was 13.8%. By 2014 last year use had fallen in the UK to 7.9%, compared to 1.2% in Hungary (Winstock 2014). At present, there is no information available online regarding mephedrone for the 2015 and 2016 sweeps of the GDS; in 2017 lifetime use in the UK was down to 1.9%; no information is given for the UK in respect of last year use, but the lowest percentage given for any country was 0.7%, suggesting that the UK rate was lower than that (Winstock et al. 2017). Levels of mephedrone use in the USA appear to be much lower than in the UK. For example, in 2012 last year use was reported at only 2%, in 2013 lifetime use had risen to 5.3% with last year use at <5% (Winstock 2012, 2013).

Other characteristics presented in the Mixmag/Global Drug Surveys include: MDPV first emerged in the 2011 sweep with lifetime use reported at 4.4% and last year use at 3.0%; a year later use in the previous 12 months had fallen to 0.5% (Winstock 2012). In 2012, 70% reported snorting mephedrone, and 20% swallowed the substance. The following year the proportion of snorting had increased to 86% and that swallowing had fallen to 12% (Winstock 2012, 2013). Nearly half (49%) of respondents in 2012 thought the purity of mephedrone had fallen compared to the previous year; its value for money being rated 5.8 out of 10 in 2013 (Winstock 2012, 2013). In 2013, the typical price for 1 g of mephedrone was £20, although 47% of UK respondents typically paid less; prior to mephedrone being controlled in April 2010 the price for 1 g was about £10. Over half of the respondents (56%) typically used 0.25–1 g in a normal session; 10% reported normally using >2 g, whilst one-third (33%) had consumed >2 g in a single session and 11% >7 g in a single session (Winstock 2013).

9.3.3 Additional United Kingdom Data

According to the 2016/2017 sweep of the Crime Survey for England and Wales (CSEW), lifetime use of mephedrone by 16–59 year olds rose from 1.9% in 2012/2013 to 2.3% the following year but has now fallen to 1.8%: rates for 16–24 year olds over the same period rose from 4.5 to 6.3% before declining to 3.3%. The rates for last year use have fallen for the wider age group from 1.3% in 2010/2011 to 0.1% in 2016/2017; amongst young people the fall was from 4.1 to 0.3%

(Broadfield 2017). These rates represent a fall from an estimated peak of 211,000 reporting use in the year prior to the survey in 2013/2014 to only 48,000 in 2016/2017. Combining the results from the 2013/2014 and 2014/2015 sweeps of the CSEW shows that mephedrone was the most likely drug to be used simultaneously with other drugs (68%), compared to ecstasy (57%) and amphetamines (50%) (Lader 2015).

In Ireland and Northern Ireland, NPS use among persons aged 15–64 years dropped significantly between 2010/2011 and 2014/2015 according to the fourth drug prevalence survey of households in Ireland and Northern Ireland (National Advisory Committee on Drugs and Alcohol & Department of Health Northern Ireland 2016). In Northern Ireland specifically, lifetime use of mephedrone rose from 2.0 to 2.5% between 2010/2011 and 2014/2015, whilst last year use fell from 1.1 to 0.5%, but last month use rose from 0.1 to 0.3% in the same period. Mephedrone usage rates were higher for males than for females in both periods: lifetime 3.1 and 3.6% for males versus 0.9 and 1.4%; last year 1.9 and 0.9% versus 0.3% in both years; last month 0.1 and 0.5% versus 0.0 and 0.1%. There were higher rates for those aged 15–34 years compared to those aged 35–64 years: lifetime 4.3 and 5.1% versus 0.4 and 0.7%; last year 2.2 and 1.2% versus 0.3 and 0.1%; last month 0.1 and 0.5% versus 0.0 and 0.1%. The declines noted both for NPS and also for mephedrone in Northern Ireland occurred during a period during which national controls on NPS were introduced by both governments and government agencies were raising the awareness of drug users to the health risks associated with NPS.

Both the general household surveys and the Mixmag/Global Drug Surveys have demonstrated a fall in the use of mephedrone in recent years, although the timing is different according to the source, e.g. 2011 in the Mixmag/Global Drug Surveys but 2013/2014 according to the CSEW, and 2014/2015 in Northern Ireland. These patterns may represent a demonstration of how the diffusion of certain drugs may first appear and peak in specific sub-populations, such as the gay community, before spreading to the dance music scene and then spreading to the wider community.

It is noticeable, that levels of mephedrone use, in particular, were highest within the gay community, even after the UK government introduced controls. Mephedrone emerged from near obscurity to become the most popular illegal drug in a survey of 308 clients of two South London gay dance clubs in July 2010 (Measham et al. 2011). Just over half (52%) of respondents had used mephedrone in the last year, 41% had taken it in the last month and 27% had either taken and/or were planning to take it on the fieldwork nights. The second-generation cathinone naphyrone ('NRG-1') had been taken by 11% of the sample in the last month and 3% had taken and/or planned to take it on the fieldwork night. Only cocaine was used more frequently than mephedrone; last year use was 59% and last month use was 44%. The majority (75–80% depending on the form) of last month ecstasy users had also used mephedrone during this period, as had two-thirds of cocaine users. Rather than displacing or replacing such stimulants, mephedrone became part of the club drug reportoire, reflecting the trend over recent years towards poly-substance use (Moore et al. 2013). These findings are replicated by another study looking at drug use in similar venues a year later: mephedrone had the highest prevalence of last month use (53.2%) and use on the night of the survey (41.0%); this compares with both cocaine (44.6 and 16.7%, respectively) and MDMA/ecstasy (26.9 and 5.8%, respectively). Other 'legal highs' were used 'on the night' to a more limited extent: methoxetamine (1.6%) and 1-benzylpiperazine, BZP (0.6%), synthetic cannabinomimetics—Spice/K2 (0.6%) and pipradrols (0.6%) (Wood et al. 2012).

In France, research into the chemsex phenomenon found that across 13 Addictovigilance centres between 2008 and 2017, synthetic cathinones occupied 7 out of the 12 psychoactive substances used, in descending order: 4-MEC (61); 3-MMC (48); mephedrone (47); MDPV (17); methylone (4); 4-P (mix of 3-MMC and 4-MEC) (3); and α-PVP (2). This compares to cocaine (52), GHB/γ-butyrolactone (GBL) (45), methylamphetamine (20), ketamine (11) and MDMA (6); most of these are stimulants. The relative proportions of cathinones in the Addictovigilance database for 2008–2017 were: 4-MEC (34%); 3-MMC (26%); mephedrone (26%); MDPV (9%); methylone (2%); 4-P (2%); α-PVP (1%). Mentions recorded by the national French Helpline during 2014–2017 were: 4-MMC/mephedrone (49%); 3-MMC (22%); 4-MEC (16%); MDPV (9%); methylone (2%); 2-MMC (2%); and α-PVP (1%) (Djezzar et al. 2017).

More generally, we can summarise that recent studies indicate that mephedrone, and most likely other synthetic cathinones, have been added to the established repertoires of psychostimulants instead of replacing or displacing ecstasy and cocaine (Schifano et al. 2017).

9.4 Investigating Motivations for Using Psychoactive Substances

Understanding motivations for drug use at an individual level helps to provide information on what factors might drive change(s) in the drug market(s) including the impact of policy changes (Reuter and Pardo 2017). Such insight can facilitate effective harm-reduction initiatives, and help to anticipate what psychoactive substances may (re-)emerge in the future (Sutherland et al. 2017).

There are a number of ways in which we can gain an insight into the opinions, beliefs, desires and motivations of individuals. At a population level, we have exploited the use of various types of surveys, with a range of theoretical and practical challenges, to see what we can understand about the nature and extent of synthetic cathinone use. However, these patterns are an aggregation and cumulation of multiple and complex decisions by individual members of communities and societies. Reports from medical settings and poison information centres do not provide information about reasons for consumers starting use of these molecules (Ashrafioun et al. 2016).

For an insight into what factors may influence and shape these decisions, as well as the nexi within which they are made, and thus try and understand such processes, it is necessary to consider individual experiences and lives. This could be done through a variety of ethnographic approaches: from individual interviews and focus groups, where access to the target population is straightforward, to using more indirect approaches as netnography (Kozinets 1998). Here, methods such as conducting thematic analyses of discussions in online drug user fora, postings on social media resources such as YouTube, Facebook, Twitter, Instagram, etc., as well as experiences collated by databases such as Erowid (https://www.erowid.org) can be employed (Davey et al. 2012; Deluca et al. 2012). For some at-risk groups, other types of survey have to be used, e.g. online surveys advertised on user fora. These can also be used to find out about what people are using, how and when, why and where, effects (desired and unwanted) using a netnographic approach (Orsolini et al. 2015a). User fora on the open Web and the DeepWeb can not only be used to investigate such as aspects but also can be used to find out, for example about prices, purity, availability and suppliers (Corkery et al. 2017).

Recently, more advanced techniques, such as social network analysis and analytic tools developed for 'big data', have been introduced from the related field of digital ethnography. For example, using 20,000+ drug experiences reported on Erowid, Krieg et al. (2017) argue for the integration of computational and digital data analyses with traditional ethnographic methods. The authors state that their study provided useful insights into the interconnections and relationships across several domains, e.g. drug phenomenology, consumption and harm reduction. This deeper understanding at the collective individual level helps researchers to make sense of developments at the wider social level (Kozinets 2015; Krieg et al. 2017).

Triangulation of information is important as this makes research findings more robust, reliable and generalisable.

9.4.1 Motivations for Use of Any Drug

One can draw a distinction between motivations (which may include more subjective influences) and reasons (which may be more objective, logical, rational and determined, in part, by external influences) for the use of psychoactive substances.

The reasons for an individual using a psychoactive substance can be straightforward and singular or complex and multidimensional. To take alcohol as an example: on the surface, a consumer may consume an alcoholic drink just to enjoy its effects. However, if one starts to consider the context of its use, a wider range of possible influences may be seen as influencing its consumption, e.g. the effects may be more apparent in a specific setting, with certain other individuals, at a particular time, etc. In turn, these may be influenced by perceived needs to 'fit in' with or 'belong' to a particular social group (i.e. to have some sort of 'identity'), to socialise, to not feel isolated, etc.

Dimensions along which motivations (including desires, wants and needs) for drug use can be described could include:

- Religious/spiritual—as part of a religious ritual (to celebrate 'sacraments' as in 'taking communion' using wine or marijuana); to be 'cleansed' or 'purified' (e.g. ayahuasca); to accompany reading a Holy Book; to enhance 'enlightenment', to commune more directly with deities or other spiritual forces;
- Exploratory/experimentation/curiosity—this can range from scientists (such as the late research chemist Alexander Shulgin) designing new molecules, self-experimenting with them and recording their effects, to 'e-Psychonauts' who intentionally experience drug-induced altered states of consciousness so as to try investigating their minds, and possibly address spiritual questions, through such direct experiences, through to 'creatives' such as artists, designers, musicians, writers, philosophers, thinkers and even academics using psychoactive substances to be more original, innovative and 'thinking outside the box' to achieve new insights or understandings of phenomena in the world(s) around them;
- Social/cultural—being part of a social (including religious) group will lead to expectations on the part of others that individuals will conform to particular norms, practices and customs (e.g. peer pressure), thereby confirming their allegiance or belonging to that group; symbolic evidence of this relationship may have to be demonstrated through the consumption of a psychoactive substance, such as at an initiation ceremony, or a 'rite of passage'. Regular consumption, at regular intervals and settings, may help to reinforce this sense of belonging, e.g. 'going out with the boys/girls' to the local bar or pub on a Friday or Saturday night. Loyalty to a group, a nation or its leader may be customary at formal occasions, e.g.'toasting the bride and groom' or 'toasting the Queen';
- Recreational—drug use is often associated with particular types of music scene, with specific drug classes being linked to individual genres of music, e.g. 'dance club' drugs; these scenes and their associated drug repertoires can be fluid and evolve;
- Employment/occupational—it may be necessary to taste psychoactive substances as part of one's occupation, for example wine-taster, brewer, whisky-blender or law enforcement agent; to consume it as a religious leader or practitioner (e.g. priest or shaman) during a religious service/rite, e.g. a 'sacrament' such as 'communion' or performing an exorcism; to facilitate business transactions;
- Therapeutic—this could be preventative or to cure (e.g. ibogaine to cure addiction to drugs), to deal with symptoms of a disease or mental health problem (whether prescribed by a health professional or self-medicated), to enhance the effects of another psychoactive substance, to ameliorate the side-effects of another psychoactive substance e.g. during 'come-down' or to facilitate withdrawal;

- Addiction/dependence—physiological/chemical dependence as well as psychological dependence on, 'craving' for a substance may develop; the abrupt cessation of use may be potentially life threatening in some extreme cases;
- Functional—searching for energy (Cadet-Taïrou 2016); to be alert/stay awake; to stave off hunger, suppress appetite, and weight-loss; passing time/avoiding boredom, etc. (Van Hout and Hearne 2015).

These dimensions may well overlap, the boundaries can be blurred and intersect with one another, having different significances as the drug using career/history of an individual evolves. The themes above will be echoed in the following sections.

The reasons for using NPS and in what contexts are likely to reflect those for illegal drugs: socio-cultural, religious, and recreational use; therapeutic/self-medication; functional goals; and exploratory/experimentation/curiosity. Typical reasons for using 'legal highs' include curiosity, pleasure and to get 'a better high than illegal drugs' (Norman et al. 2014). Initial experimentation and continued use may also be driven by such factors as boredom and peer socialisation (Van Hout and Hearne 2015).

Factors specifically encouraging NPS use include: they are comparatively easier to source; their lower price; and (assumed) better quality and safety compared with traditional drugs. For individuals and groups subject to drug screening, such as prisoners, armed forces, law enforcement agencies, hospital in-patients, individuals undergoing drug treatment or detoxification, drivers of vehicles, trains or aircraft, and the assumed, and actual, non-detectability in tests of NPS can be an important factor. However, the creation and continual evolution of the Internet and social media appear to effect NPS use, especially 'legal highs', in additional ways and for additional reasons, including: no stigma being attached to their use; they are easily available; they can be purchased anonymously online; and there is no need to have a prescription for medications.

The development of the so-called 'Drugs 2.0' Internet has helped to facilitate a new drug 'scenario', where a new generation of users possessing quasi-expert pharmacological and chemical knowledge and ideas regarding NPS and/or combinations of NPS and known molecules share information via virtual platforms (Power 2013; Schifano et al. 2006). Such e-Psychonauts also sell and buy NPS and other novel substances through this new cybermarket (Orsolini et al. 2015b; Power 2013, Schifano et al. 2006). These self-experimenting psychonauts historically typically investigated the '*inner universe*'/'*Psychocosmos*' (Jünger 1970), often through psychotropic ('*turning towards the psyche*') drugs (Labate and Jungaberle 2011). These days, they could be referred to as pharmacological 'trippers' (Corkery et al. in press), since, as we saw earlier, they are individuals who deliberately experience states of altered consciousness, that are drug induced to try and explore their minds, and find answers to spiritual questions via direct experiences (Carroll 1987; Labate and Jungaberle 2011; Power 2013).

Drug use by psychonauts can be regarded as somewhat similar to ritual plant consumption by ancient and current shamans from an anthropological point of view, as they appear to favour mostly NPS with hallucinogenic/entheogenic properties (Orsolini et al. 2016, 2017). However, some consumers of NPS see a clear distinction between the neoshamanic practices of psychonauts' mental exploration and genuine shaman practice that is healing focused (Power 2013). Like shamans, e-Psychonauts may be conceived as a 'sub-cultural virtual group', and similarly, their chief beliefs and goals often have more to do with the achievement of altered states of consciousness rather than with recreational purposes (Psychonauts.com 2017; Reddit.com 2017). It could be argued (Corkery et al. in press) that the cultural reputation and attraction of NPS mainly derives from their use among e-Psychonauts (Deluca et al. 2012), whose extensive information and experience sharing on the Internet facilitates the diffusion of the 'psychonaut cultural phenotype' (Psychonauts.com 2017).

Earlier, we observed that the context or setting of drug use can influence what is taken. Taking the example of music, Cadet-Taïrou (2016) notes the association of phenethylamines with the 'electro' music scene in France. A range of external factors may influence NPS use in recreational night-life settings: the lower the age and educational level of users the more likely they were to consume NPS in a private setting (home of the user or a friend) rather than in a public setting such as a party or other event; occupational status and whether the consumer is the main salary earner in a household can also influence use of NPS.

As with traditional recreational drugs, temporal patterns can be observed in the consumption of NPS (Corkery et al. 2015). For example, α-PVP, BZP, mephedrone and methylone were mostly used at weekends, similar to the pattern reported for cocaine and MDMA, whilst opioids and cannabis were used at a consistent level through the course of the week, in a study of wastewater samples analysed in Adelaide, Australia, between 2011 and 2015 (Tscharke et al. 2016).

It was noted earlier that NPS use is higher in some groups compared to the general population, including: young people; people in contact with mental health services; the homeless; people who inject drugs (PWID); and men who have sex with men (MSM) (UNODC 2017a). A recent Scottish study found that amongst such groups, synthetic cannabinoids were used by 41% of subjects and mephedrone by 19% (MacLeod et al. 2016). Marginalised groups, including the young, socially disadvantaged and homeless with little disposable income, are more likely to be attracted to NPS as they are cheaper but have similar effects (MacLeod et al. 2016; NDEWS 2016).

Economic factors (drivers, pushers) such as demand and supply, have been associated with the emergence of NPS in relation to the market for traditional recreational drugs, and the subsequent establishment and evolution of the NPS market (Tscharke et al. 2016; Van Hout and Hearne 2015). Eight years ago the main drivers for NPS use were the poor quality of traditional drugs, especially cocaine, amphetamines and ecstasy/MDMA, and their lack of availability and high price (Corkery et al. in press; Guirguis et al. 2017); ease of access on line, 'value for money', rather than their apparent safety (assumed because of their 'quasi-legal' status) are now the impelling forces, as the quality of such traditional drugs has now recovered to (Winstock et al. 2016; MacLeod et al. 2016) or exceeded their former levels.

9.4.1.1 Reasons Why Individuals Use Synthetic Cathinones

There is considerable overlap between the reasons for using khat (Corkery 2016; and see Chap. 3) and for using synthetic cathinones. A range of reasons are reported, including: hallucinogenic experiences, euphoria, mood enhancement, openness in communication, empathy, mental clarity, increased alertness, insomnia, stimulation, intensification of sensory experiences, increased energy, reduced appetite, increased sociability, increased libido and sexual performance (Rosenbaum et al. 2012; Schifano et al. 2016), increased confidence (Brookman et al. 2017). 'Problem reduction' is also reported, e.g. mephedrone being taken to alleviate psychological problems and to help deal with emotional issues (Brookman et al. 2017).

Motivations for use of synthetic cathinones have varied over time, in different countries and amongst different user groups. These differences may be due, in part, to the varying patterns in the pace at which the knowledge of and use of synthetic cathinones have spread. For example, mephedrone use first emerged in 'sentinel' groups such as the 'gay' community in London before going mainstream in the UK and other European countries, before the bath salts craze hit the USA. By this time, MDPV was entering the scene and thus the North American experience was not identical to the European one. The highest number of calls to national poison information centres in the USA happened more than a year after the peak in the UK (Spyker et al. 2012).

Increased availability, ease of purchase via the Internet, perceived legality, assumed high purity, perceived better high, fewer side-effects, safer to use and short duration of action have all been cited as reasons for using synthetic cathinones by a sub-sample of ecstasy users in Australia in 2014 (Sutherland et al. 2017). In previous years, surveys of and interviews with users suggested that the increased popularity and convenience of purchasing mephedrone was associated with the unavailability, high price and impurity of several stimulants, chiefly amphetamine, cocaine and MDMA (ACMD 2010; Carhart-Harris et al. 2011; Measham et al. 2010).

Data collected between April and September 2014 in Hungary from 198 opioid substitution therapy clients showed that these users substituted heroin for NPS, mainly cathinones, because of practical issues in obtaining and using the drugs rather than a preference of their psychopharmacological effects. In decreasing order of significance, these reasons included: curiosity, replacing other drugs, increased availability, because their friends used it, more intense subjective effects, low price, legal status, poor detectability, exotic brand name, perceived safety, shorter effect duration and attractive packaging. Patients who shifted and substituted their opioid dependence to cathinones favoured mephedrone, methylone, 4-MEC and MDPV (Bretteville-Jensen et al. 2013; Kapitány-Fövény et al. 2017). Brettville-Jensen et al. (2013) added that perceived lack of long- or short-term harm, good ratings from peers or on the Internet, good-quality substitutes of low purity cocaine and ecstasy at the time, lack of detection, with an effect that mimics that of ecstasy and other club drugs, are other motivations for use of cathinones. Other practical reasons put forward by users include the range of methods by which they can be used (Brookman et al. 2017).

Medical practitioners, and users, report that the high from mephedrone lasts longer than that from heroin, requires fewer 'hits', can lessen the withdrawal effects of heroin, and may assist in harm-reduction through users migrating from heroin and crack cocaine (with their associated addiction and withdrawal problems) to mephedrone (Brookman et al. 2017).

In the context of chemsex, Djezzar et al. (2017) report the reasons for cathinone use, in descending order of importance, as: increasing sexual pleasure (>80%); alter consciousness (c. 70%); hard sex (c. 70%); disinhibition (c. 60%); atypical sexual behaviour (c. 45%); prolonging erection (<10%).

9.4.1.2 Consumption with Other Substances

Combining the use of synthetic cathinones with the other classes of psychoactive drugs, particularly stimulants, is very common and to be expected in today's poly-pharmacy recreational drug scene(s), as are counteracting classes such as downers (sedatives, opiates/opioids, benzodiazepines, GHB/GBL, and dissociatives). These drug classes can include prescription drugs, anaesthetic agents and alcohol (Valente et al. 2014).

Consumption of synthetic cathinones with other psychoactive substances may occur since, as with other drug types, they may counteract the side-effects (e.g. taking alcohol after consuming mephedrone to reverse its side-effects), or prolong the desired-for effects of a molecule, or enhance it in some other way, e.g. make it more potent. Different combinations may be used in different settings and for different purposes. For example, mephedrone may be consumed during partying alongside cocaine, MDMA, GBL and ketamine (Kirby and Thomber-Dunwell 2013), whereas in 'chem-sex' sessions MDMA, GBL and methamphetamine may be co-ingested with mephedrone (Bourne et al. 2015). Such combinations induce intensified sexual experiences and increased disinhibition (e.g. leading to risky behaviours, including not using protection during sex, and having multiple partners) (Batisse et al. 2016; Dargan et al. 2011; Karila et al. 2015).

Experiences shared on Bluelight and Drugs-Forum between 2010 and 2015 included: mephedrone being used with Viagra (Cialis) to boost libido and improve sexual performance (e.g. prolonging erection); 4-MEC with 'weed' and GBL during 'chemsex' sessions (Drugs-Forum 2010a, b, 2015); MDPV with cocaine; amphetamines and methamphetamine for their entactogenic effects; MDPV with β-blockers to counteract tachycardia; MDPV with GHB/GBL, and 5-methoxy-*N*-methyl-*N*-isopropyltryptamine (5-MeO-MiPT) to improve libido; MDPV with zopiclone to enhance visual hallucinations; caffeine for its stimulant effect; famotidine, omeprazole and domperidone to counteract stomach pain; and benzodiazepines for their anxiolytic effects (Coppola and Mondola 2012; Katz et al. 2014).

Amphetamines, cocaine and ecstasy tablets may be used singly or in combination with cathinones to maintain arousal and a state of alertness, since MDMA/

serotonergic stimulant and entactogenic effects wear off after a few hours (Schifano 2004). Theoretically, potential interactions occur between cathinones and prescribed medications. For example, the monoamine oxidase inhibitor Moclobemide can be taken to enhance the effects of ecstasy-like stimulants (Vuori et al. 2003). These are selective serotonin reuptake inhibitors (SSRIs) that could interact to induce serotonin syndrome (Green et al. 1995), which can be fatal (Schifano 2004). Such an outcome can occur when serotonergic release is intensified by the parallel use of dopaminergic stimulants, such as cocaine and amphetamines (Huether et al. 1997; Schifano 2004). Ketamine and/or LSD together with stimulants/cathinones can increase the likelihood of idiosyncratic/intoxicated 'behaviour', such as dangerous driving, with a consequent higher risk of a lethal outcome. Concurrent use of sedatives could ameliorate excess sympathomimetic overactivity observed with stimulant use, including cathinones (Schifano 2004).

Reasons for using certain psychoactive substances are arguably wider than purely intentional motivations, as in the sense of a desire, or more strongly a craving for consuming them. Use of synthetic cathinones may be symptomatic of wider influences at work. For example, a representative sample of 30,377 US adolescents from the 8th, 10th and 12th grade participating in the Monitoring the Future study from 2012 to 2014 showed that motivation for use of both synthetic cannabinomimetics and synthetic cathinones was correlated with truancy, cigarette smoking, the use of alcohol and marijuana (Patrick et al. 2016).

9.4.1.3 Comparison of Synthetic Cathinones with Other Stimulants

Whilst some cathinones have behavioural effects, respectively, similar to amphetamine or methylamphetamine and cocaine, others produce effects that are closer to those of MDMA (see Chap. 6). Typically, four categories of synthetic cathinones are described (Guirguis et al. 2017):

- Cocaine-MDMA-mixed, e.g. mephedrone, 4-MEC, methylone, etylone, butylone and naphyrone. Such molecules are substrates for the dopamine transporter (DAT), serotonin transporter (SERT), and norepinephrine transporter (NET). When ingested orally these molecules are associated with an entactogenic, MDMA-like, effect, but when administered intranasally they have a psychostimulant cocaine-like effect (Liechti 2015);
- Methamphetamine-like, e.g. cathinone, methcathinone, flephedrone, ethcathinone and 3-fluoromethcathinone. Such molecules are monoamine transporter substrates with DAT selective profiles, exhibit high inhibitory potencies at DAT and show lower inhibitory potencies at SERT (Simmler et al. 2013, 2014). Similar to methylamphetamine, they induce DA and NE release (Liechti 2015);
- MDMA-like, e.g. methedrone and 4-trifluoromethylmethcathinone. Such molecules show a greater inhibitory potency at SERT compared to their inhibition of DAT, whilst inducing release of both NE and 5-HT-like amphetamine analogues

such as MDMA, paramethoxymethamphetamine (PMMA), paramethoxyamphetamine (PMA), 4-ethylthioamphetamine (4-MTA) (Simmler et al. 2014).

- Pyrovalerones, e.g. pyrovalerone, MDPV and α-PVP. Such molecules are non-substrate transporter inhibitors, exhibiting inhibitory potencies at DAT and NET equal or greater than cocaine (Baumann et al. 2013) or methylamphetamine (Aarde et al. 2013). Both MDPV and α-PVP are regarded as cocaine-like molecules (Smith et al. 2017). However, recent research suggests that the pyrovalerone analogue α-PVT possesses rewarding and reinforcing effects akin to both cocaine and methylamphetamine (Cheong et al. 2017).

Some users reported that a cathinone-induced stimulant effect is comparable to methylphenidate at low doses or to a combined effect of both amphetamine and cocaine at high doses (Coppola and Mondola 2012). Others consider cathinones to be pharmacologically similar to and as potent as amphetamine, cocaine, and MDMA (EMCDDA 2012).

The 'honeymoon' period for mephedrone lasted for a few years. It slowly became apparent that compared to cocaine and MDMA, mephedrone had some less favourable effects. Winstock (2012) reports that users ceased their 'love affair' with mephedrone, in part, due to its higher scores for extreme agitation, headaches, tremors, nausea and feeling depressed after use, and to a lesser extent paranoia and chest pain. In terms of positive effects of using drugs, mephedrone came second out of ten drugs for increased pleasure from social interactions, but had more negative effects, i.e. first for not being able to function normally in the days after use, negative effects on mental health; second for negative effects on ability to work/study/progress, negative effects on physical health; third for unpleasant physical and psychological effects when intoxicated (Winstock 2013).

9.4.1.4 Use of Specific Types of Synthetic Cathinones

Typically, the literature classifies cathinones according to their pharmacological action and in comparison with more traditional stimulant drugs. However, there is no consensus on how to do this, or the categories to be used. Some approaches look at effects in terms of different substrates or non-substrate transporter inhibitors (Simmler et al. 2013). Others draw comparisons with certain 'classical' classes of traditional drugs of abuse, i.e. MDMA, amphetamine/methylamphetamine, cocaine (Concheiro et al. 2013) or even combinations such as MDMA/cocaine, together with chemical structure categories (Assi et al. 2017; Concheiro et al. 2013; Valente et al. 2014). Some of these approaches also single out pyrrolidines as a separate group. These approaches lead to overlapping of categories and blurred distinctions.

Attempting to establish a connection between different types of cathinones based on their DAT/SERT inhibition ratio, structural characteristics or other parameters and their similarity to traditional stimulants is difficult in the context of non-psychonaut consumers. An ideal solution in terms of categorising synthetic cathinones would be to employ a single criterion in terms of their similarity to the effects of 'classical' drugs of abuse. Therefore, a pragmatic approach has been adopted for a study being conducted by the lead author and colleagues into motivations for using specific types of synthetic cathinones, i.e. similarity to the effects of these classical drugs or combinations thereof, and the pyrrolidine group: amphetamine/methylamphetamine-like; cocaine-like; MDMA-like; MDMA- + cocaine-like; and pyrrolidines/pyrovalerone-type. However, there are problems with ascribing particular effects to specific cathinone molecules or particular classes or families of cathinones and asking users about them. The wording of questions has to be very clear as to which of these is actually being asked about. It is unlikely that a non-psychonaut user will appreciate the distinction. It may even be the case that only very few psychonauts will feel it necessary to distinguish in this way what they consume.

The number of papers which have come closest to achieving this is very small, i.e.: Winstock et al. (2011a), Cahart-Harris et al. (2011), Johnson and Johnson (2014), and Ashrafioun et al. (2016). The first paper demonstrated that a web-based survey recruiting from an existing survey (Mixmag) could be employed to study users' experiences, but only contained a limited number of questions exploring motivations for use of particular types of synthetic cathinones . The second study also used a web-based survey, recruiting through advertisements on user fora, to ask about specific subject effects but did not investigate reasons for use. The third study followed a similar recruitment strategy to the second one and did get some information on reasons for using 'bath salts' as a generic class of drug: curiosity; mind/brain exploration; liking the effect; to remain alert/awake; to avoid failing drug tests; to enhance sexual experiences. The fourth study listed above, set out to overcome the limitations of the previous three, using online recruitment through advertisements on specific websites; put a recruitment video in YouTube and purchased targeted advertisements on Facebook (Ashrafioun et al. 2016).

The commonest reasons, in decreasing order for first trying synthetic cathinones were: curiosity, believing incorrectly that they were taking another substance; social pressure; having been offered 'bath salts'. The main reasons given for consuming 'bath salts' were: fun/enjoyment/euphoria/rush; curiosity/new experience; substitution for another drug; being at a party/event. The authors note that the effects and motivations echo those who use other illegal stimulants, and are similar to those of the previous three studies.

The co-consumption of several synthetic cathinones may occur. If a drug user consumes a branded product they may unintentionally take a number of different synthetic cathinones at the same time, as suppliers may seek to use up old stocks of molecules which have subsequently become regulated or controlled, to test the market with 'new' molecules, or to produce several different effects. Many of the structural pyrovalerone analogues that replaced MDPV in the cathinones market possess similar mechanisms of action and pharmacological effects. As these molecules were offered for sale with the same brand names, this fact was not evident to buyers or to consumers (Beck et al. 2016). During the period January 2011–March 2016, the STRIDA project in Sweden recorded 11 pyrovalerone drugs in addition to MDPV and α-PVP across 114 intoxicated patients with 149 total pyrovalerone-type findings; only 8 patients had used a single cathinone, one patient had used 5 pyrovalerone molecules in combination. Co-administration of pyrovalerones and/or use in combination with amphetamines suggests that rather than using a specific pyrovalerone, patients were principally interested in obtaining a central stimulatory effect (Beck et al. 2017). Discussions on online user fora also suggest that as well as suggesting particular combinations of cathinones, some suggest combining 4-MEC with the amphetamine-type 4-FA (Van Hout 2014).

In addition to this, adulterants and cutting agents may be added to counteract side-effects, or since they have a similar chemical structure to index molecules they may add to the complexity of the mixture, and have a stronger/masking detection signal hindering identification (Guirguis et al. 2017). On the other hand, some branded products may contain undisclosed synthetic cathinones; this could be very dangerous for consumers.

Reasons for the deliberate consumption of several cathinones may include, one could conjecture, a desire to self-harm or to attempt suicide (perhaps as ‘a cry for help’ or as a serious but unsuccessful attempt to kill oneself). Such a reason may be due to the paranoia or depression induced by synthetic cathinones (Corkery et al. 2012). Another reason for co-administering cathinones, may be a belief that taking two or more synthetic cathinones simultaneously can cause a synergy between them, leading to an enhanced experience (Iversen et al. 2014). Rat studies suggest that the co-administration of MDPV, mephedrone, and methylone may intensify their associated neurochemical effects via a sustained increase in DA and serotonin (5-HT) in various brain regions through an additive effect (Allen-Campbell et al. 2016). Binary mixtures of common ‘bath salts’ constituents (i.e. MDPV, methylone, and caffeine) are generally additive in nature; however, both supra-additive (MDPV + caffeine and methylone + caffeine) and sub-additive interactions (MDPV + methylone) can be observed (Collins et al. 2017).

9.5 Stopping Use of Cathinones

As was the case with reasons for initiating and continuing consumption of synthetic cathinones, so with stopping the use of them, in terms of drawing a distinction between motivations and reasons for doing so. Here again, the possibility of overlap and reinforcement does exist.

Dealing first with motivations, tolerance, dependence and withdrawal symptoms have been reported by many synthetic cathinone users. However, there is little-published information to date on tolerance, craving, dependence/addiction and withdrawal in respect of these aspects (Prosser and Nelson 2012). Ross and Peselow (2009) suggested that ‘the dopaminergic stimulation of the reward system could explain the development’ of these consequences. These factors can create difficulties in ceasing cathinone use. In respect of mephedrone, users have reported tolerance, craving (being ‘moreish’), need to re-dose, loss of control, addiction and stimulant withdrawal syndrome; echoing those found amongst amphetamine and cocaine-dependent individuals (Bajaj et al. 2010; Brookman et al. 2017; Carhart-Harris et al. 2011; Dargan et al. 2010; EMCDDA 2011; Reed 2010; Winstock et al. 2011b). The evidence is lacking on these issues in respect of methylone (Karila et al. 2016). For MDPV there is some limited animal evidence for some of these features (tolerance, craving, dependence and withdrawal syndrome) (Coppola and Mondola 2012; Gannon et al. 2017). The structural parent of MDPV, α-PVP appears to give rise to extreme craving and risk of binge consumption (EMCDDA 2016b).

Unpleasant physical and mental symptoms may be experienced when going through withdrawal from synthetic cathinones. Feelings of depression and anxiety following use of mephedrone have been reported, but no physical withdrawal syndrome has been described (Dargan et al. 2011; Prosser and Nelson 2012). Tiredness, insomnia, nasal congestion and impaired concentration have also been described as symptoms indicating mephedrone withdrawal (Winstock et al. 2011b). Psychological dependency is possible in terms of mephedrone (Dargan et al. 2011). Following the abrupt stopping of chronic mephedrone, methcathinone and MDPV, some users have reported symptoms of anxiety and depression, craving, anhedonia, anergia and insomnia (Centers for Disease Control and Prevention (CDC) 2011; Winstock et al. 2011a).

Whilst Van Hout (2014) notes that 4-MEC users did not report withdrawal symptoms akin to those of mephedrone (i.e. depression, anergia, anhedonia or insomnia), there is a case report of neonatal withdrawal syndrome in a baby whose mother was a chronic 4-MEC user 'The newborn presented with increased jitteriness and irritability, high pitched cry, hypertonia in the limbs and brisk tendon reflexes' (Pichini et al. 2014). In terms of treatment for cathinone withdrawal, bupropion has been suggested (Coppola and Mondola 2012; Lev-Ran 2012).

Other motivations for the cessation of use may include: previously experienced health harms, including negative impacts on physical and mental well-being, e.g. personality change, as well as not liking consuming NPS (MacLeod et al. 2016), including cathinones. The most common adverse reactions due to cathinones use are restlessness and anxiety, ranging from mild agitation to severe psychosis. Furthermore, tachycardia, hypertension, abdominal pain, chills, flushing, sweating, hyperthermia, renal failure, rhabdomyolysis and seizures can also be observed (Schifano et al. 2017).

Reasons for discontinuing use can include social pressures, e.g. peer pressure. Negative effects on relationships with family and friends could act as an incentive to change (Ashrafioun et al. 2016; Brookman and Bennett 2017; Brookman et al. 2017). Additional experiences which may also influence change include: recognising that one has neglected responsibilities, missing school or work, feeling that use is a problem, realising that money spent on buying cathinones could have been used for other purposes, shame/embarrassment, avoidance by others (Ashrafioun et al. 2016).

Additional economic factors that could influence reduced consumption of synthetic cathinones or even cessation of use are the corollaries of those influences which helped push their use in 2009–2011: increased price, reduced availability or reduced quality of cathinones, increased availability and quality of amphetamines, cocaine and ecstasy/MDMA.

Such external reasons or drivers such as those briefly outlined above are more likely to be influenced by policy and treatment, thereby making individuals more receptive/susceptible to change.

Major reasons for encouraging individuals to stop their use of synthetic cathinones include the potentially very serious physiological and psychiatric/psychological consequences that may occur. These, as a result of chronic use of both natural and synthetic cathinones, can include: suicidal ideation, attempted suicide, as well as completed suicide; paranoia, including audio and visual hallucinations; psychoses; depression; mood disturbances; sleep deprivation; loss of appetite leading to rapid weight loss (Brookman et al. 2017). Other reasons to stop can include: the avoidance of acquisitive crime to fund use, fear of acquiring a criminal record and/or being imprisoned, reducing the potential for violence when under the influence, debt avoidance, avoiding loss of employment and/or being made homeless (Brookman and Bennett 2017; Brookman et al. 2017).

9.6 Injecting Behaviours and Blood-Borne Infections

The injecting of NPS is of increasing concern in Europe, the USA and elsewhere. There are several aspects to this issue, which will be briefly described, as the consequences of injecting synthetic cathinones can be very serious and injectors/users need to be aware of/educated about these in the hope that they will make rational decisions in relation to these risks.

Injecting drugs, especially powders or crushed pills (the commonest form of synthetic cathinones) diluted in a liquid, can cause damage to the injection sites, venous system and cause blockages such as thrombi—with associated risks of death. Injecting stimulant users are likely to inject more frequently due to their relatively short length of action, and sharing contaminated injecting equipment, thereby increasing health risks (Fischer et al. 2013; UNODC 2017a).

At-risk populations include subgroups of Men having sex with Men (MSM), 'long-term abstinent former opiate users, people who inject other drugs such as heroin and amphetamines, people who switch to injecting synthetic cathinones and people who have switched from snorting to injecting' (UNODC 2017a). Such changes have occurred in Hungary during 2011–2015 (Tarján et al. 2017). Typical synthetic cathinones injected by PWID in Hungary changed over time: 2010 mephedrone; 2011 MDPV and 4-MEC; 2012–2013 pentedrone; 2014 α-PVP; 2015 α-pyrrolidinohexiophenone (α-PHP) (Hungerian Focal Point 2015; Péterfi et al. 2014; Rácz et al. 2015).

Within the chemsex context in the UK, synthetic cathinones/mephedrone misuse, either on their own or together with methamphetamine and/or GHB/GBL, especially in the MSM sub-population, has been associated with unprotected sex practices, sharing syringes, and highly elevated risk of the spread of blood-borne and sexually transmitted diseases including HIV and hepatitis C (Bourne et al. 2014; 2015; UNODC 2017a). HIV rates amongst synthetic cathinone injectors appear to be higher than in 'traditional' drug-injecting populations, e.g. mephedrone, MDPV and methylone injectors in San Diego, California (Wagner et al. 2014), in Ireland amongst homeless α-PVP injectors who often reused syringes and shared filters (Giese et al. 2015), English mephedrone injectors (PHE 2015), and pentedrone injectors in Hungary (Rácz et al. 2016).

9.7 Morbidity and Mortality

A detailed description of the adverse medical consequences, including death, resulting from the use of synthetic cathinones is outside the scope of this chapter. However, it is important for (would-be) users of such molecules to be aware of the potential risks to which they may expose themselves. This may influence their decisions to use in the first place, make changes in their practice(s), or even stop.

9.7.1 Intoxications, Hospital Admissions

Cathinones are also characterised by increased presentations to emergency departments for both physical and mental health associated toxicity (UNODC 2017a). For example, during 2011 there were 22,904 visits to US Emergency Departments (ED) recorded by the Drug Abuse Warning Network (DAWN) as caused by bath salts. This represented about 0.9% of all ED visits. One-third of cases involved solely bath salts, 15% were implicated with natural cannabis or synthetic cannabinoids, and 52% with combinations of other drugs (DAWN 2013). In Florida, between September 2014 and December 2015 an outbreak of α-PVP use was associated with 80 deaths and thousands of ED admissions, many from excited delirium syndrome—hyperstimulation, paranoia and hallucinations that can lead to violent aggression and self-injury (NDEWS 2016).

The Euro-DEN project compiled information on 10,709 ED admissions across 16 centres in 10 European cities for the period 2014–2015. NPS accounted for 7.3% of admissions, of which 61.2% were cathinones (mephedrone 73.9% of these). There were 9 deaths related to NPS out of a total of 67; all of these involved cathinones (Dargan 2017). A retrospective study conducted in Germany over a 6-year period between January 2010 and January 2016 on cathinone self-reported users presenting to ED due to cathinone-related complications revealed that 64% of patients (n = 81) were males with a median age of 34 years (Romanek et al. 2017). In this study, cathinones which related to higher rates of intoxications were MDPV, methylone and 3-MMC, in reducing order of frequency (Romanek et al. 2017).

9.7.2 *Fatalities*

A range of synthetic cathinones (including mephedrone, methylone and butylone, ethylone, α-PVP, MDPV and methedrone) has been associated with fatalities (Schifano et al. 2017). This is especially true in the UK, where the number of synthetic cathinone-related poisoning deaths rose from 6 in 2010 to a peak of 49 in 2015, falling to 25 in 2016, with a cumulative total of 157. The majority (122) of these involved mephedrone (ONS 2015, 2016a, b, c, 2017a, b). In Scotland, the number of cathinone deaths registered rose from 1 in 2009 to 8 in 2010, falling to 6 in 2014 and 2015, and down to 1 in 2016, with a cumulative total of 32. The majority (17) of these involved mephedrone (NRS 2017). In Northern Ireland, the number of cathinone deaths registered rose from 1 in 2010 to 3 before falling to 1 in 2016, with a cumulative total of 27. Only 6 of these involved mephedrone (NISRA 2016; and personal communication to lead author for provisional 2016 data). In summary, since 2009 the cumulative total of synthetic cathinone-related poisoning deaths registered in the UK by the end of 2016 was 216, of which 145 involved mephedrone.

In a 3-year review of NPS detected in 203 in-life or post-mortem blood and urine and in UK criminal case works, mephedrone and 4-MEC were the most common NPS identified. Interestingly, all fatalities related to cathinones in this review were hangings and mechanical suicides (Elliott and Evans 2014). Another 3-year review conducted in Poland, from 2012 to 2014, for 112 cases where NPS were identified in biological material from people driving under the influence of drugs, cathinones were prevalent in 88% of the cases. In decreasing frequency, these were: 3-MMC, α-PVP, pentedrone, 3′,4′-methylenedioxy-α-pyrrolidinobutyrophenone (MDPBP), ethcathinone, mephedrone, MDPV, 4-MEC, buphedrone and methylone (Adamowicz et al. 2016).

9.8 What Challenges Might the Future Hold?

The different constituencies dealing with NPS are likely to continue facing challenges which may further compound the problem of getting a comprehensive understanding of synthetic cathinones. Further new synthetic cathinone molecules are likely to appear, leading to not only intoxications but also to psychoses, paranoia and other psychiatric issues. Deaths are also likely to continue, especially against a background of poly substance consumption.

There are reports of 'vaping' equipment being used to administer NPS/'legal highs' (Jones 2016; Thurtle et al. 2017). There is some evidence of this occurring with mexedrone (Roberts et al. 2017), so it is conceivable that users of other cathinones could experiment with such a method of administration.

Following the introduction of 'blanket ban' legislation in countries such as Ireland and the UK, and disappearance of 'smart' or 'head' shops, it is likely that the Web (Surface and Deep) will increasingly utlise to source supplies of synthetic cathinones. NPS are now being traded alongside traditional 'street' drugs such heroin, cocaine and cannabis in clandestine street drug markets (Shapiro and Daly 2017).

9.9 Conclusions

This chapter has looked at the current and emerging knowledge and challenges of synthetic cathinones from epidemiological, anthropological and clinical pharmacological perspectives. With so much remaining unknown about a wide range of issues concerned with NPS, including synthetic cathinones, the need for inter-disciplinary approaches to the ever-evolving NPS phenomenon is clearly demonstrable. The need for timely, up to date information and the development of methods to understand the potential characteristics and properties of new molecules still remains. Forensic as well as socio-anthropological approaches are required to have a deeper understanding of synthetic cathinones and the implications of their use. This will help policy-makers, educationalists, health and other professionals, service providers and funders to develop responses including prevention and intervention initiatives. Furthermore, (would-be) psychoactive substance users can be assisted in making more rational decisions about synthetic cathinones.

The key dimensions that need documenting are the human ones of why people are looking for new synthetic cathinones, amongst other available NPS, what effects they are seeking, what they are actually taking (how, how much, where and when), and what the consequences of their actions may be. Furthermore, external influences or factors which could play a causative role in how these play out also need to be understood. The use of historic or traditional methods of epidemiology can play a part, but new and innovative approaches need to be continually improved and developed in a world of increasing reliance on rapid electronic exchange and dissemination of information.

Acknowledgements The following provided unpublished data on mephedrone which were extracted for this study: Home Office Crime and Policing Analysis Unit; Justice Analytical Services, Scottish Government; and PSNI Statistics Branch.

This chapter was supported in part by grants of the European Commission (Drug Prevention and Information Programme 2014–16; contract no. JUST/2013/DPIP/AG/4823; EU-MADNESS project). Further financial support was provided by the EU Commission-targeted call on cross border law enforcement cooperation in the field of drug trafficking—DG Justice/DG Migrations and Home Affairs (JUST/2013/ISEC/DRUGS/AG/6429) Project EPS/NPS (Enhancing Police Skills concerning Novel Psychoactive Substances; NPS).

Some of this material has been presented as an oral presentation:

John M. Corkery (presenter), Amira Guirguis, Laura Orsolini, Duccio Papanti and Fabrizio Schifano (2017). Oral presentation. An investigation into the relationship(s) between the different chemical classes of synthetic cathinones and their effects: desired, adverse and toxic. Fifth International Conference on Novel Psychoactive substances. United Nations Office on Drugs and Crime (UNODC) Vienna International Centre, Vienna 23–24 October 2017.

Conflicts of Interest
The authors are unaware of any potential conflicts of interest. However, F.S. is a full member of the UK Advisory Council on the Misuse of Drugs (ACMD) and its NPS Committee; J.C. was a member of the ACMD's Working Groups on Drug-related deaths (1999–2000 and 2016–7), and is currently a co-opted member of the Technical Committee (2016 to date) and NPS Committee (2009 to date). JC was responsible for producing drug statistics for the Home Office (1994–2002), acted as the UK Focal Point on Drugs' expert on drug-related deaths and mortality related to drug use (2000–2015), and also contributed to the UK Annual Report Questionnaire to the UNODC over the period 1994–2013. The views expressed here reflect only the authors' views and not necessarily those of the Home Office or the ACMD.

References

Aarde SM, Huang PK, Creehan KM et al (2013) The novel recreational drug 3,4-methylenedioxypyrovalerone (MDPV) is a potent psychomotor stimulant: self-administration and locomotor activity in rats. Neuropharmacology 71:130–140. https://doi.org/10.1016/j.neuropharm.2013.04.003

ACMD (2010) Consideration of the cathinones. In: Advisory Council on the Misuse of Drugs. London: Home Office. https://www.gov.uk/government/uploads/system/uploads/attachment_data/file/119173/acmd-cathinodes-report-2010.pdf. Accessed 13 Nov 2017

Adamowicz P, Gieroń J, Gil D et al (2016) The prevalence of new psychoactive substances in biological material—a three-year review of casework in Poland. Drug Test Anal 8(1):63–70. https://doi.org/10.1002/dta.1924

Allen-Campbell SA, Oakes HV, Rednour SA et al (2016) Neurochemical outcomes following individual and combined synthetic cathinone exposure. FASEB J 30(1):Suppl. 1186.6

Ashrafioun L, Bonadio FA, Baik KD et al (2016) Patterns of use, acute subjective experiences, and motivations for using synthetic cathinones ("Bath Salts") in recreational users. J Psychoactive Drugs 48(5):336–343. https://doi.org/10.1080/02791072.2016.1229875

Assi S, Gulyamova N, Kneller P, Osselton D (2017) The effects and toxicity of cathinones from the users' perspectives: a qualitative study. Hum Psychopharmacol 32(3). https://doi.org/10.1002/hup.2610

Bajaj N, Mullen D, Wylie S (2010) Dependence and psychosis with 4-methylmethcathinone (mephedrone) use. BMJ Case Rep 3, pii: bcr0220102780. https://doi.org/10.1136/bcr.02.2010.2780

Batisse A, Grégoire M, Marillier M et al (2016) Usage de cathinones à Paris [Cathinones use in Paris]. L'Encéphale 42(4):354–360. https://doi.org/10.1016/j.encep.2015.09.002

Baumann MH, Partilla JS, Lehner KR et al (2013) Powerful cocaine-like actions of 3,4-methylenedioxypyrovalerone (MDPV), a principal constituent of psychoactive 'bath salts' products. Neuropsychopharmacology 38(4):552–562. https://doi.org/10.1038/npp.2012.204

Beck O, Franzén L, Bäckberg M et al (2016) Toxicity evaluation of α-pyrrolidinovalerophenone (α-PVP): results from intoxication cases within the STRIDA project. Clin Toxicol (Phila) 54(7):568–575. https://doi.org/10.1080/15563650.2016.1190979

Beck O, Bäckberg M, Signell P, Helander A (2017) Intoxications in the STRIDA project involving a panorama of psychostimulant pyrovalerone derivatives, MDPV copycats. Clin Toxicol (Phila). https://doi.org/10.1080/15563650.2017.1370097

Bourne A, Reid D, Hickson F et al (2014) The chemsex study: drug use in sexual settings among gay and bisexual men in Lambeth, Southwark & Lewisham. Sigma Research, London School of Hygiene & Tropical Medicine, London. https://www.lambeth.gov.uk/sites/default/files/ssh-chemsex-study-final-main-report.pdf. Accessed 13 Nov 2017

Bourne A, Reid D, Hickson F et al (2015) Illicit drug use in sexual settings ('chemsex') and HIV/STI transmission risk behaviour among gay men in South London: findings from a qualitative study. Sex Transm Infect 91(8):564–568. https://doi.org/10.1136/sextrans-2015-052052

Bretteville-Jensen AL, Tuv SS, Bilgrei OR et al (2013) Synthetic cannabinoids and cathinones: Prevalence and markets. Forensic Sci Rev 25(1–2):7–26

Broadfield D (ed.) (2017) Drug misuse: findings from the 2016/17 Crime Survey for England and Wales. Statistical bulletin 11/17. 27July. Home Office, London. https://www.gov.uk/government/uploads/system/uploads/attachment_data/file/642738/drug-misuse-2017-hosb1117.pdf, https://www.gov.uk/government/uploads/system/uploads/attachment_data/file/633263/drug-misuse-1617-tables.xlsx. Accessed 12 Oct 2017

Brookman F, Bennett TH (2017) Users' accounts of the links between mephedrone use and violence. Eur J Criminol. https://doi.org/10.1177/1477370816686121

Brookman F, Bennett TH, Hills R (2017) The pleasures and pains of mephedrone use; perceptions of users and practitioners. Drugs educ prev policy 24(1):103–110. https://doi.org/10.1080/09687637.2016.1192106

Cadet-Taïrou A (2016) New psychoactive substances: user profiles and practices. Tendances No. 108, 8 p (Apr 2016)

Carhart-Harris RL, King LA, Nutt DJ (2011) A web-based survey on mephedrone. Drug Alcohol Depend 118(1):19–22. https://doi.org/10.1016/j.drugalcdep.2011.02.011

Carroll PJ (ed) (1987) Liber null & psychonaut: an introduction to chaos magic. Samueal Weiser Inc., York Beach, ME

Centers for Disease Control and Prevention (CDC) (2011) Emergency department visits after use of a drug sold as "bath salts"—Michigan, 13 November, 2010–March 31, 2011. MMWR Morb Mortal Wkly Rep 60(19):624–627

Cheong JH, Choi MJ, Jang CG et al (2017) Behavioral evidence for the abuse potential of the novel synthetic cathinone alpha-pyrrolidinopentiothiophenone (PVT) in rodents. Psychopharmacology (Berl) 234(5):857–867. https://doi.org/10.1007/s00213-017-4526-8

Collins GT, Gannon BM, Galindo KI et al (2017) Abuse-related effects of "Bath Salts" mixtures: studies with MDPV, methylone, and caffeine in rats. FASEB J 31(1):Suppl 987.8

Concheiro M, Anizan S, Ellefsen K, Huestis MA (2013) Simultaneous quantification of 28 synthetic cathinones and metabolites in urine by liquid chromatography-high resolution mass spectrometry. Anal Bioanal Chem 405(29):9437–9448. https://doi.org/10.1007/s00216-013-7386-z

Coppola M, Mondola R (2012) Synthetic cathinones: chemistry, pharmacology and toxicology of a new class of designer drugs of abuse marketed as "bath salts" or "plant food". Toxicol Lett 211(2):144–149. https://doi.org/10.1016/j.toxlet.2012.03.009

Corkery JM (2016) Khat—chewing it over: continuing 'cultural cement', cardiac challenge or catalyst for change? In: Davies S, Johnston A, Holt D (eds) Forensic toxicology—drug use and misuse. Royal Society of Chemistry, London, pp 166–207

Corkery JM, Schifano F, Ghodse AH (2012) Mephedrone-related fatalities in the United Kingdom: contextual, clinical and practical issues. In: Gallelli L (ed) Pharmacology. InTech—Open Access Publisher, Rijeka, Croatia, pp 355–380. https://doi.org/10.5772/32935

Corkery JM, Loi B, Claridge H et al (2015) Gamma hydroxybutyrate (GHB), gamma butyrolactone (GBL) and 1,4-butanediol (1,4-BD; BDO): a literature review with a focus on UK fatalities related to non-medical use. Neurosci Biobehav Rev 53:52–78. https://doi.org/10.1016/j.neubiorev.2015.03.012

Corkery JM, Orsolini L, Papanti GD, Schifano F (2017) From concept(ion) to life after death/the grave: the 'natural' history and life-cycle(s) of novel psychoactive substances (NPS). Hum Psychopharmacol. https://doi.org/10.1002/hup.2566

Corkery JM, Orsolini L, Papanti D, Schifano F (in press) Novel psychoactive substances (NPS) and recent scenarios: epidemiological, anthropological and clinical pharmacological issues. In: Miolo G, Stair JL, Zloh M (eds) Light in forensic science: issues and applications. Royal Society of Chemistry, London (Chapter 9)

Dargan PI (2017) The Euro-DEN Plus project—use of a sentinel centre model to collect data on acute drug and new psychoactive substance (NPS) toxicity in Europe. Oral presentation. In: Fifth International Conference on Novel Psychoactive Substances. United Nations Office on Drugs & Crime, Vienna International Centre, Austria, 23–24 Oct 2017
Dargan PI, Albert S, Wood DM (2010) Mephedrone use and associated adverse effects in school and college/university students before the UK legislation change. QJM 103(11):875–879. https://doi.org/10.1093/qjmed/hcq134
Dargan PI, Sedefov R, Gallegos A, Wood DM (2011) The pharmacology and toxicology of the synthetic cathinone mephedrone (4-methylmethcathinone). Drug Test Anal 3(7–8):454–463. https://doi.org/10.1002/dta.312
Davey Z, Schifano F, Corazza O, Deluca P, Psychonaut Web Mapping Group (2012) e-Psychonauts: conducting research in online drug forum communities. J Ment Health 21(4):386–394. https://doi.org/10.3109/09638237.2012.682265
DAWN (2013) "Bath Salts" were involved in over 20,000 drug-related Emergency Department visits in 2011. The DAWN Report, 17 September. Drug Abuse Warning Network. https://www.samhsa.gov/data/sites/default/files/spot117-bath-salts-2013/spot117-bath-salts-2013.pdf. Accessed 5 Oct 2017
DEA (2012) US Drug Enforcement Administration Office of Diversion Control. National Forensic Laboratory Information System: year 2011 annual report. In: Administration USDE (ed), Springfield, VA. https://www.deadiversion.usdoj.gov/nflis/2011annual_rpt.pdf#search=2011%20annual%20report%20nflis. Accessed 13 Nov 2017
DEA (2016) Emerging Threat Report Annual 2016. US Drug Enforcement Administration, Springfield, VA. https://ndews.umd.edu/sites/ndews.umd.edu/files/emerging-threat-report-2016-annual.pdf. Accessed 27 Oct 2017
DEA (2017a) Emerging Threat Report Mid-year 2017. US Drug Enforcement Administration, Springfield, VA. https://ndews.umd.edu/sites/ndews.umd.edu/files/dea-emerging-threat-report-2017-mid-year.pdf. Accessed 27 Oct 2017
DEA (2017b) Emerging Threat Report Third Quarter 2017. US Drug Enforcement Administration, Springfield, VA. https://ndews.umd.edu/sites/ndews.umd.edu/files/emerging-threat-report-2017-quarter3.pdf. Accessed 27 Oct 2017
DEA (2017c) 2017 National Drug Threat Assessment, 23 October. US Drug Enforcement Administration, Springfield, VA. https://www.dea.gov/docs/DIR-040-17_2017-NDTA.pdf. Accessed 27 October 2017
Deniker P, Lôo H, Cuche H, Roux JM (1975) Utilisation abusive par les toxicomanes d'un psycho-stimulant, la pyrovalérone [Abuse of pyrovalerone by drug addicts]. Ann Med Psychol (Paris) 2(4):745–748
Deluca P, Davey Z, Corazza O et al (2012) Identifying emerging trends in recreational drug use; outcomes from the Psychonaut Web Mapping Project. Prog Neuropsychopharmacol Biol Psychiatry 39(2):221–226. https://doi.org/10.1016/j.pnpbp.2012.07.011
Djezzar S, Batisse A, Marillier M, Chevallier C (2017) Chemsex in France through Addictovigilance Network tools and Researches. Oral presentation. In: Fifth International Conference on Novel Psychoactve Substances. United Nations Office on Drugs & Crime, Vienna International Centre, Austria, 23–24 Oct 2017
Drugs-Forum (2010a) Mephedrone + Viagra (or Cialis)? https://drugs-forum.com/forum/showthread.php?t=96002. Accessed 1 Dec 2016
Drugs-Forum (2010b) Mephedrone + Viagra to boost libido. https://drugs-forum.com/forum/showthread.php?t=96002. Accessed 1 Dec 2016
Drugs-Forum (2015) 4-MEC + weed. https://drugs-forum.com/forum/showthread.php?t=279054. Accessed 1 Dec 2016
Drummond A, Codd M, Donnelly N et al (2014) Study on the prevalence of drug use, including intravenous drug use, and blood-borne viruses among the Irish prisoner population. National Advisory Committee on Drugs and Alcohol, Dublin
ECCD (1985) WHO Expert Committee on drug dependence. 22nd meeting. World Health Organization Technical Report Series, 729

Elliott S, Evans J (2014) A 3-year review of new psychoactive substances in casework. Forensic Sci Int 243:55–60. https://doi.org/10.1016/j.forsciint.2014.04.017

EMCDDA (2011) Report on the risk assessment of mephedrone in the framework of the Council Decision on new psychoactive substances. https://doi.org/10.2810/40800. http://www.emcdda.europa.eu/system/files/publications/571/TDAK11001ENC_WEB-OPTIMISED_FILE_280269.pdf_en. Accessed 18 Oct 2017

EMCDDA (2012) The state of the drugs problem in Europe. Annual report 2012. https://doi.org/10.2810/64775. http://www.emcdda.europa.eu/attachements.cfm/att_190854_EN_TDAC12001ENC_.pdf. Accessed 13 Nov 2017

EMCDDA (2016a) EU Drug market report. In-depth analysis. http://www.emcdda.europa.eu/system/files/publications/2373/TD0216072ENN.PDF. Accessed 13 Nov 2017

EMCDDA (2016b) Report on the risk assessment of 1-phenyl-2-(pyrrolidin-1-yl)pentan-1-one (α-pyrrolidinovalerophenone, α-PVP). https://doi.org/10.2810/71700. http://www.emcdda.europa.eu/system/files/publications/2934/TDAK16001ENN.pdf Accessed 6 Nov 2017

EMCDDA (2017) European Drug Report: Trends and Developments 2017. https://doi.org/10.2810/0610791. http://www.emcdda.europa.eu/system/files/publications/4541/TDAT17001ENN.pdf_en. Accessed 1 Oct 2017

EMCDDA–Europol (2017) 2016 Annual Report on the implementation of Council Decision 2005/387/JHA. http://www.emcdda.europa.eu/system/files/publications/4724/TDAN17001ENN_PDFWEB.pdf_en. Accessed 24 Sept 2017

Erowid. https://www.erowid.org. Accessed 15 Oct 2017

Fischer A, Curruthers S, Power R et al (2013) The link between amphetamine-type stimulant use and the transmission of HIV and other blood-borne viruses in the Southeast Asia region. ANCD research paper no. 25. National Drug Research Institute, Australian National Council on Drugs, Melbourne. http://www.leahn.org/wp-content/uploads/2013/12/rp25-amphetamine-type-stimulants-ANCD-2012.pdf. Accessed 13 Nov 2017

Gannon BM, Rice KC, Collins GT (2017) Reinforcing effects of abused 'bath salts' constituents 3,4-methylenedioxypyrovalerone and α-pyrrolidinopentiophenone and their enantiomers. Behav Pharmacol 28(7):578–581. https://doi.org/10.1097/FBP.0000000000000315

Gardos G, Cole JO (1971) Evaluation of pyrovalerone in chronically fatigued volunteers. Curr Ther Res Clin Exp 13(10):631–635

Giese C, Igoe D, Gibbons Z et al., outbreak control team (2015) Injection of new psychoactive substance snow blow associated with recently acquired HIV infections among homeless people who inject drugs in Dublin, Ireland, 2015. Euro Surveill 20(40). https://doi.org/10.2807/1560-7917.es.2015.20.40.30036

Green AR, Cross AJ, Goodwin GM (1995) Review of the pharmacology and clinical pharmacology of 3,4-methylenedioxymethamphetamine (MDMA or "Ecstasy"). Psychopharmacology (Berl) 119(3):247–260

Guirguis A, Corkery JM, Stair JL et al (2017) Intended and unintended use of cathinone mixtures. Hum Psychopharmacol 32(3). https://doi.org/10.1002/hup.2598

Huether G, Zhou D, Rüther E (1997) Causes and consequences of the loss of serotonergic presynapses elicited by the consumption of 3,4-methylenedioxymethamphetamine (MDMA, "ecstasy") and its congeners. J Neural Transm (Vienna) 104(8–9):771–794

Hungerian Focal Point (2015) 2014 National Report (2013 data) to the EMCDDA by the Reitox National Focal Point: Hungary (Budapest, 2015). http://www.emcdda.europa.eu/system/files/publications/1006/HU_National_Report_2014_en.pdf. Accessed 13 Nov 2017

Iversen L, White M, Treble R (2014) Designer psychostimulants: pharmacology and differences. Neuropharmacology 87:59–65. https://doi.org/10.1016/j.neuropharm.2014.01.015

Johnson PS, Johnson MW (2014) Investigation of "bath salts" use patterns within an online sample of users in the United States. J Psychoactive Drugs 46(5):369–378. https://doi.org/10.1080/02791072.2014.962717

Jones I (2016) High on vapour. New Scientist 232(3104–4106):21. https://doi.org/10.1016/S0262-4079(16)32322-3

Jünger E (1970) Annäherungen: Drogen und Rausch [Approaches: drugs and intoxication]. Verlag IL Kunst, Köln, NRW, Germany

Kapitány-Fövény M, Farkas J, Pataki PA et al (2017) Novel psychoactive substance use among treatment-seeking opiate users: the role of life events and psychiatric symptoms. Hum Psychopharmacol 32(3). https://doi.org/10.1002/hup.2602

Karila L, Megarbane B, Cottencin O, Lejoyeux M (2015) Synthetic cathinones: a new public health problem. Curr Neuropharmacol 13(1):12–20. https://doi.org/10.2174/1570159X13666141210224137

Karila L, Billieux J, Benyamina A et al (2016) The effects and risks associated to mephedrone and methylone in humans: a review of the preliminary evidences. Brain Res Bull 126(Pt 1):61–67. https://doi.org/10.1016/j.brainresbull.2016.03.005

Karila L, Lafaye G, Scocard A et al (2017) MDPV and α-PVP use in humans: the twisted sisters. Neuropharmacology pii: S0028–3908(17)30474–4. https://doi.org/10.1016/j.neuropharm.2017.10.007

Katz DP, Bhattacharya D, Bhattacharya S et al (2014) Synthetic cathinones: "a khat and mouse game". Toxicol Lett 229(2):349–356. https://doi.org/10.1016/j.toxlet.2014.06.020

Kirby T, Thomber-Dunwell M (2013) High-risk drug practices tighten grip on London gay scene. Lancet 381(9861):101–102

Kozinets RV (1998).On netnography: Initial reflections on consumer research investigations of cyberculture. Adv Consumer Res 25:366–371. http://acrwebsite.org/volumes/8180/volumes/v25/NA-25. Accessed 8 Oct 2017

Kozinets RV (ed) (2015) Netnography: redefined, 2nd edn. Sage Publications Ltd., London

Krieg LJ, Berning M, Hardon A (2017) Anthropology with algorithms? An exploration of online drug knowledge using digital methods. Med Anthropol Theory 4(3):21–52. https://doi.org/10.17157/mat.4.3.458

Labate BC, Jungaberle H (eds) (2011) The internationalization of Ayahuasca. Lit Verlag, Zurich, Switzerland

Lader D (ed) (2015) Drug misuse: findings from the 2014/15 crime survey for England and Wales 2nd ed. (July) Home Office, London. https://www.gov.uk/government/uploads/system/uploads/attachment_data/file/462885/drug-misuse-1415.pdf. Accessed 12 Oct 2017

Lev-Ran S (2012) A case of treating cathinone dependence and comorbid depression using bupropion. J Psychoactive Drugs 44(5):434–436. https://doi.org/10.1080/02791072.2012.736851

Liechti ME (2015) Novel psychoactive substances (designer drugs): overview and pharmacology of modulators of monoamine signalling. Swiss Med Wkly 14 Jan, 145, w14043. https://doi.org/10.4414/smw.2015.14043

MacLeod K, Pickering L, Gannon M et al (2016) Understanding the patterns of use, motives, and harms of new psychoactive substances in Scotland. Social Research Series, November. Scottish Government, Edinburgh. http://www.gov.scot/Resource/0051/00510607.pdf. Accessed 13 Nov 2017

Matthews A, Bruno R (2010) Mephedrone use among regular ecstasy consumers in Australia. EDRS Drug Trends Bulletin, December 2010. National Drug and Alcohol Research Centre, University of New South Wales, Sydney. https://ndarc.med.unsw.edu.au/sites/default/files/ndarc/resources/EDRS%20Bulletin%20Dec%202010.pdf. Accessed 8 Oct 2017

McNamara S, Stokes S, Coleman N (2010) Head shop compound abuse amongst attendees of The Drug Treatment Centre Board. The Drug Treatment Centre Board, Dublin. http://www.drugsandalcohol.ie/13185/1/Headshopcompounts.pdf. Accessed 17 Oct 2017

Measham F, Moore K, Newcombe R (2010) Tweaking, bombing, dabbing and stockpiling: the emergence of mephedrone and the perversity of prohibition. Drugs Alcohol Today 10(1): 14–21. https://doi.org/10.5042/daat.2010.0123

Measham F, Wood DM, Dargan PI, Moore K (2011) The rise in legal highs: prevalence and patterns in the use of illegal drugs and first- and second-generation "legal highs" in South London gay dance clubs. J Subst Use 16(4):263–272. https://doi.org/10.3109/14659891.2011.594704

Meyers K, Kaynak Ö, Bresani E et al (2015) The availability and depiction of synthetic cathinones (bath salts) on the internet: do online suppliers employ features to maximize purchases? Int J Drug Policy 26(7):670–674. https://doi.org/10.1016/j.drugpo.2015.01.012

Moore K, Dargan PI, Wood DM, Measham F (2013) Do novel psychoactive substances displace established club drugs, supplement them or act as drugs of initiation? The relationship between mephedrone, ecstasy and cocaine. Eur Addict Res 19(5):276–282. https://doi.org/10.1159/000346678

National Advisory Committee on Drugs and Alcohol & Department of Health Northern Ireland (2016) Prevalence of drug use and gambling in Ireland and drug use in Northern Ireland. Bulletin No. 1. National Advisory Committee on Drugs and Alcohol, Dublin. http://health.gov.ie/wp-content/uploads/2016/11/Bulletin-1.pdf. Accessed 13 Oct 2017

National Drug Early Warning System (NDEWS), Coordinating Center (2016) Southeastern Florida (Miami Area), Sentinel Community Site (SCS), Drug use patterns and trends, 2016. http://www.miamidade.gov/advocacy/library/florida-substance-abuse-report.pdf. Accessed 13 Oct 2017

NHS Digital (2017) Smoking, drinking and drug use among young people in England—2016. Statistics Team, The Health and Social Care Information Centre. ISBN: 978-1-78734-116-6. https://digital.nhs.uk/media/33663/Smoking-Drinking-and-Drug-Use-Among-Young-People-in-England-2016-Report/default/sdd-2016-rep. Tables—drug use prevalence, England 2016 (Chapter 9). http://digital.nhs.uk/pubs/sdd2016. Accessed 5 Nov 2017

NISRA (2016) Drug-related deaths and deaths due to drug misuse registered in Northern Ireland (2005–2015). Northern Ireland Statistics & Research Agency, Belfast. https://www.nisra.gov.uk/sites/nisra.gov.uk/files/publications/Drug_Tables_15.xls. Accessed 19 May 2017

Norman J, Grace S, Lloyd C (2014) Legal high groups on the internet. The creation of new organized groups? Drugs educ prev policy 21(1):14–23. https://doi.org/10.3109/09687637.2013.769500

NRS (2017) Drug-related deaths in Scotland in 2016. National Records of Scotland, Edinburgh. https://www.nrscotland.gov.uk/statistics-and-data/statistics/statistics-by-theme/vital-events/deaths/drug-related-deaths-in-scotland/2016. https://www.nrscotland.gov.uk/statistics-and-data/statistics/statistics-by-theme/vital-events/deaths/drug-related-deaths-in-scotland/2016/list-of-tables-and-figures. Accessed 16 Aug 2017

OJEU (2017) Official Journal of the European Union, L305, vol 60, 21 Nov 2017. http://eur-lex.europa.eu/legal-content/EN/TXT/PDF/?uri=OJ:L:2017:305:FULL&from=EN. Accessed 23 Nov 2017

ONS (2015) Drug-related deaths registered in England and Wales in 2014. Ad hoc data request no. 005096, released 10 Dec 2015. Office for National Statistics, Newport, Gwent. https://www.ons.gov.uk/peoplepopulationandcommunity/birthsdeathsandmarriages/deaths/adhocs/005096drugrelateddeathsregisteredinenglandandwalesin2014. Accessed 19 May 2017

ONS (2016a) Drug related deaths involving new psychoactive substances, England and Wales, 2015 registrations. Ad hoc data request no. 006158, released 29 Sept 2016. Office for National Statistics, Newport, Gwent. https://www.ons.gov.uk/peoplepopulationandcommunity/healthandsocialcare/drugusealcoholandsmoking/adhocs/006158drugrelateddeathsinvolvingnewpsychoactivesubstancesenglandandwales2015registrations. Accessed 19 May 2017

ONS (2016b) Number of drug-related deaths where a new psychoactive substance was the only drug mentioned on the death certificate, England and Wales, deaths registered in 1993–2015. Ad hoc data request no. 006187, released 4 Oct 2016. Office for National Statistics, Newport, Gwent. https://www.ons.gov.uk/peoplepopulationandcommunity/healthandsocialcare/causesofdeath/adhocs/006187numberofdrugrelateddeathswhereanewpsychoactivesubstancewastheonlydrugmentionedonthedeathcertificateenglandandwalesdeathsregisteredin1993to2015. Accessed 19 May 2017

ONS (2016c) Drug-related deaths mentioning a new psychoactive substance, England and Wales, 2011 to 2015 registrations. Ad hoc data request no. 006413, released 30 Nov 2016. Office for National Statistics, Newport, Gwent. https://www.ons.gov.uk/peoplepopulationandcommunity/

birthsdeathsandmarriages/deaths/adhocs/006413drugrelateddeathsmentioninganewpsychoactivesubstanceenglandandwales2011to2015registrations. Accessed 19 May 2017

ONS (2017a) Deaths related to drug poisoning in England and Wales: 2016 registrations, 2 August. Office for National Statistics, Newport, Gwent. https://www.ons.gov.uk/peoplepopulationandcommunity/birthsdeathsandmarriages/deaths/bulletins/deathsrelatedtodrugpoisoninginenglandandwales/2016registrations/pdf, https://www.ons.gov.uk/file?uri=/peoplepopulationandcommunity/birthsdeathsandmarriages/deaths/datasets/deathsrelatedtodrugpoisoningenglandandwalesreferencetable/current/deathsrelatedtodrugpoisoningenglandandwales.xls. Accessed 6 Aug 2017

ONS (2017b) Drug-related deaths mentioning a new psychoactive substance, England and Wales, 2011 to 2015 registrations. Ad hoc data request no 007307, released 4 August 2017. Office for National Statistics, Newport, Gwent. https://www.ons.gov.uk/peoplepopulationandcommunity/birthsdeathsandmarriages/deaths/adhocs/007307deathsrelatedtodrugpoisoninginvolvingspecificsubstancesenglandandwalesdeathsregisteredin2016. https://www.ons.gov.uk/file?uri=/peoplepopulationandcommunity/birthsdeathsandmarriages/deaths/adhocs/007307deathsrelatedtodrugpoisoninginvolvingspecificsubstancesenglandandwalesdeathsregisteredin2016/2016drugrelateddeathsbyspecificsubstances.xls. Accessed 6 Aug 2017

Orsolini L, Papanti GD, Francesconi G, Schifano F (2015a) Mind navigators of chemicals' experimenters? A web-based description of e-psychonauts. Cyberpsychol Behav Soc Netw 18 (5):296–300. https://doi.org/10.1089/cyber.2014.0486

Orsolini L, Francesconi F, Papanti GD et al (2015b) Profiling the online recreational/prescription drugs' customers and overview of the drug vending virtual marketplaces. Hum Psychopharmacol 30(4):302–318. https://doi.org/10.1002/hup.2466

Orsolini L, St John-Smith P, McQueen D et al (2016) Evolutionary considerations on the emerging subculture of the e-psychonauts and the novel psychoactive substances: a comeback to the shamanism? Curr Neuropharmacology 15(5):731–737. https://doi.org/10.2174/1570159X15666161111114838

Orsolini L, Papanti D, Corkery J, Schifano F (2017) An insight into the deep web; why it matters for addiction psychiatry? Hum Psychopharmacol 32(3). https://doi.org/10.1002/hup.2573

Palamar JJ, Martins SS, Su MK, Ompad DC (2015) Self-reported use of novel psychoactive substances in a US nationally representative survey: prevalence, correlates, and a call for new survey methods to prevent underreporting. Drug Alcohol Depend 156:112–119. https://doi.org/10.1016/j.drugalcdep.2015.08.028

Palamar JJ, Acosta P, Calderón FF et al (2017) Assessing self-reported use of new psychoactive substances: the impact of gate questions. Am J Drug Alcohol Abuse 43(5):609–617. https://doi.org/10.1080/00952990.2017.1322094

Patrick ME, O'Malley PM, Kloska DD et al (2016) Novel psychoactive substance use by US adolescents: characteristics associated with use of synthetic cannabinoids and synthetic cathinones. Drug Alcohol Rev 35(5):586–950. https://doi.org/10.1111/dar.12372

Péterfi A, Tarján A, Horváth GC et al (2014) Changes in patterns of injecting drug use in Hungary: a shift to synthetic cathinones. Drug Test Anal 6(7–8):825–831. https://doi.org/10.1002/dta.1625

Public Health England (2015) Shooting up: infections among people who injected drugs in the United Kingdom, 2014 November. Public Health England, Health Protection Scotland, Public Health Wales, and Public. Health Agency Northern Ireland, London. http://www.drugsandalcohol.ie/11975/1/Shooting_Up_UK_2015.pdf. Accessed 13 Nov 2017

Pichini S, Rotolo MC, García J et al (2014) Neonatal withdrawal syndrome after chronic maternal consumption of 4-methylethcathinone. Forensic Sci Int 245:e33–e35. https://doi.org/10.1016/j.forsciint.2014.10.027

Power M (ed) (2013) Drugs 2.0. The web revolution that's changing how the world gets high. Portobello Books Ltd., London

Prosser JM, Nelson LS (2012) The toxicology of bath salts: a review of synthetic cathinones. J Med Toxicol 8(1):33–42. https://doi.org/10.1007/s13181-011-0193-z

Psychonauts.com. http://www.psychonauts.com. Accessed 15 Oct 2017

Rácz J, Csák R, Lisznyai S (2015) Transition from "old" injected drugs to mephedrone in an urban micro segregate in Budapest, Hungary: a qualitative analysis. J Substance Use 20(3):178–186. https://doi.org/10.3109/14659891.2014.895872

Rácz J, Gyarmathy VA, Csák R (2016) New cases of HIV among people who inject drugs in Hungary: false alarm or early warning? Int J Drug Policy 27:13–16. https://doi.org/10.1016/j.drugpo.2015.05.026

Reddit.com. http://www.reddit.com/r/onions/comments/22z3qe/grams_beta_version/. Accessed 15 Oct 2017

Reed J (2010) Clubbers are 'turning to new legal high mephedrone', 13 Jan. BBC Newsbeat. http://www.bbc.co.uk/newsbeat/article/10004366/clubbers-are-turning-to-new-legal-high-mephedrone. Accessed 15 Oct 2017

Reuter P, Pardo B (2017) Can new psychoactive substances be regulated effectively? An assessment of the British Psychoactive Substances Bill. Addiction 112(1):25–31. https://doi.org/10.1111/add.13439

Roberts L, Ford L, Patel N et al (2017) 11 Analytically confirmed cases of mexedrone use among polydrug users. Clin Toxicol (Phila) 55(3):181–186. https://doi.org/10.1080/15563650.2016.1271424

Romanek K, Stenzel J, Schmoll S et al (2017) Synthetic cathinones in Southern Germany—characteristics of users, substance-patterns, co-ingestions, and complications. Clin Toxicol (Phila) 55(6):573–578. https://doi.org/10.1080/15563650.2017.1301463

Rosenbaum CD, Carreiro SP, Babu KM (2012) Here today, gone tomorrow … and back again? A review of herbal marijuana alternatives (K2, Spice), synthetic cathinones (bath salts), kratom, Salvia divinorum, methoxetamine, and piperazines. J Med Toxicol 8(1):15–32. https://doi.org/10.1007/s13181-011-0202-2

Ross S, Peselow E (2009) The neurobiology of addictive disorders. Clin Neuropharmacol 32 (5):269–276

Saem de Burnaga Sanchez MJ (1929) Sur un homologue de l'éphédrine [On a homologue of ephedrine]. Bulletin de la Société Chimique de France 45:284–286

Schifano F (2004) A bitter pill. Overview of ecstasy (MDMA, MDA) related fatalities. Psychopharmacology 173(3–4):242–248. https://doi.org/10.1007/s00213-003-1730-5

Schifano F, DeLuca P, Baldacchino A et al (2006) Drugs on the web; The Psychonaut 2002 EU project. Prog Neuropsychopharmacol Biol Psychiatry 30(4):640–646. https://doi.org/10.1016/j.pnpbp.2005.11.035

Schifano F, Papanti GD, Orsolini L, Corkery JM (2016) Novel psychoactive substances: the pharmacology of stimulants and hallucinogens. Expert Rev Clin Pharmacol 9(7):943–954. https://doi.org/10.1586/17512433.2016.1167597

Schifano F, Orsolini L, Papanti D, Corkery J (2017) NPS: medical consequences associated with their intake. Curr Top Behav Neurosci 32:351–380. https://doi.org/10.1007/7854_2016_15

Schulenberg JE, Johnston LD, O'Malley PM et al (2017) Monitoring the Future national survey results on drug use, 1975–2016, vol II. College students and adults ages 19–55. Institute for Social Research, The University of Michigan, Ann Arbor. http://monitoringthefuture.org//pubs/monographs/mtf-vol2_2016.pdf. Accessed 8 Oct 2017

Shapiro H, Daly M (2017) Highways and buyways: a snapshot of UK drug scenes 2016. DrugWise, London. http://www.drugwise.org.uk/wp-content/uploads/Highwaysandbyways.pdf. Accessed 9 Feb 2017

Simmler LD, Buser TA, Donzelli M et al (2013) Pharmacological characterization of designer cathinones in vitro. Br J Pharmacol 168(2):458–470. https://doi.org/10.1111/j.1476-5381.2012.02145.x

Simmler LD, Rickli A, Hoener MC, Liechti ME (2014) Monoamine transporter and receptor interaction profiles of a new series of designer cathinones. Neuropharmacology 79:152–160. https://doi.org/10.1016/j.neuropharm.2013.11.008

Sindicich N, Burns L (2011) An overview of the 2011 EDRS: what is happening to ecstasy and related drugs in Australia. EDRS Drug Trends Bulletin, October 2011. National Drug and Alcohol Research Centre, University of New South Wales, Sydney. https://ndarc.med.unsw.

edu.au/sites/default/files/ndarc/resources/EDRS%20bulletin%20October%202011.pdf. Accessed 8 Oct 2017

Sindicich N, Burns L (2013) An overview of the 2013 ecstasy and related drugs reporting system. Drug Trends Bulletin, October. https://ndarc.med.unsw.edu.au/sites/default/files/newsevents/events/EDRS%20October%202013%20Bulletin.pdf. Accessed 8 Oct 2017

Smith DA, Blough BE, Banks ML (2017) Cocaine-like discriminative stimulus effects of amphetamine, cathinone, methamphetamine, and their 3,4-methylenedioxy analogs in male rhesus monkeys. Psychopharmacology (Berl) 234(1):117–127. https://doi.org/10.1007/s00213-016-4444-1

Spyker DA, Thomas S, Bateman DN et al (2012) International trends in designer amphetamine abuse in UK and US, 2009–2012. Clin Toxicol (Phila) 50(7):636, Abstract 141. 2012 Annual meeting of the North American Congress of Clinical Toxicology (NACCT), 1–6 Oct 2012, Las Vegas, Nevada. https://doi.org/10.3109/15563650.2012.700015. http://www.tandfonline.com/doi/pdf/10.3109/15563650.2012.700015. Accessed 29 Oct 2017

Stogner JM, Miller BL (2013) Investigating the 'bath salt' panic: the rarity of synthetic cathinone use among students in the United States. Drug Alcohol Rev 32(5):545–549. https://doi.org/10.1111/dar.12055

Sutherland R, Peacock A, Whittaker E et al (2016) New psychoactive substance use among regular psychostimulant users in Australia, 2010–2015. Drug Alcohol Depend 61:110–118. https://doi.org/10.1016/j.drugalcdep.2016.01.024

Sutherland R, Bruno R, Peacock A et al (2017) Motivations for new psychoactive substance use among regular psychostimulant users in Australia. Int J Drug Policy 43:23–32. https://doi.org/10.1016/j.drugpo.2016.12.021

Tarján A, Dudás M, Wiessing L et al (2017) HCV prevalence and risk behaviours among injectors of new psychoactive substances in a risk environment in Hungary—an expanding public health burden. Int J Drug Policy 41:1–7. https://doi.org/10.1016/j.drugpo.2016.11.006

Thurtle N, Abouchedid R, Archer JR et al (2017) Prevalence of use of electronic nicotine delivery systems (ENDS) to vape recreational drugs by club patrons in South London. J Med Toxicol 13(1):61–65. https://doi.org/10.1007/s13181-016-0583-3

Tscharke BJ, Chen C, Gerber JP, White JM (2016) Temporal trends in drug use in Adelaide, South Australia by wastewater analysis. Sci Total Environ 565:384–391. https://doi.org/10.1016/j.scitotenv.2016.04.183

UNODC (2014) Global synthetic drugs assessment 2014: amphetamine-type stimulants and new psychoactive substances. United Nations Office on Drugs and Crime, Vienna. https://www.unodc.org/documents/scientific/2014_Global_Synthetic_Drugs_Assessment_web.pdf. Accessed 10 July 2015

UNODC (2016a) World Drug Report 2016. United Nations Office on Drugs and Crime, Vienna. http://www.unodc.org/doc/wdr2016/WORLD_DRUG_REPORT_2016_web.pdf. Accessed 1 Oct 2017

UNODC (2016b) Global SMART update 2016, vol 16. United Nations Office on Drugs and Crime, Vienna. https://www.unodc.org/documents/scientific/Global-SMART-Update-2016-vol-16.pdf. Accessed 1 Oct 2017

UNODC (2017a) World drug report 2017, vol 4. Market analysis of synthetic drugs: Amphetamine-type stimulants, new psychoactive substances. United Nations Office on Drugs and Crime, Vienna. https://www.unodc.org/wdr2017/field/Booklet_4_ATSNPS.pdf. Accessed 12 Sept 2017

UNODC (2017b) Global synthetic drugs assessment: amphetamine-type stimulants and new psychoactive substances, 24 Oct. United Nations Office on Drugs and Crime, Vienna. https://www.unodc.org/documents/scientific/Global_Drugs_Assessment_2017.pdf. Accessed 27 Oct 2017

Valente MJ, Guedes de Pinho P, de Lourdes Bastos M et al (2014) Khat and synthetic cathinones: a review. Arch Toxicol 88(1):15–45. https://doi.org/10.1007/s00204-013-1163-9

Van Hout MC (2014) An internet study of user's experiences of the synthetic cathinone 4-methylethcathinone (4-MEC). J Psychoactive Drugs 46(4):273–286. https://doi.org/10.1080/02791072.2014.934979

Van Hout MC, Hearne E (2015) A community-based study of synthetic cannabinoid use in Co. Monaghan, Ireland, July. Teach na Daoine Family Resource Centre, Monaghan, Ireland. https://www.oireachtas.ie/parliament/media/committees/healthandchildren/health2015/Report-on-Synthetic-Cannabinoid-use-in-Monaghan-Town.pdf. Accessed 13 Nov 2017

Vento AE, Schifano F, Gentili F et al (2013) Bupropion perceived as a stimulant by two patients with a previous history of cocaine misuse. Ann Ist Super Sanita 49(4):402–405. https://doi.org/10.4415/ANN_13_04_14

Vuori E, Henry JA, Ojanperä I et al (2003) Death following ingestion of MDMA (ecstasy) and moclobemide. Addiction 98(3):365–368

Wadsworth E, Drummond C, Kimergård A, Deluca P (2017) A market on both "sides" of the law: the use of the hidden web for the sale of new psychoactive substances. Hum Psychopharmacol 32(3). https://doi.org/10.1002/hup.2596

Wagner KD, Armenta RF, Roth AM et al (2014) Use of synthetic cathinones and cannabimimetics among injection drug users in San Diego, California. Drug Alcohol Depend 141:99–106. https://doi.org/10.1016/j.drugalcdep.2014.05.007

Winstock A (2012) The Mixmag Drugs Survey, Mixmag Apr 2012, 5p. https://issuu.com/mixmagfashion/docs/drugs_survey_2012_2. Accessed 13 Oct 2017

Winstock A (2013) Mixmag/Global Drug Survey. Mixmag, May 2013, 3p. https://issuu.com/mixmagfashion/docs/mm_may13_drug_survey. Accessed 13 Oct 2017

Winstock A (2014) The Global Drug Survey 2014 findings. https://www.globaldrugsurvey.com/wp-content/uploads/2014/04/last-12-months-drug-prevalence.pdf. Accessed 13 Oct 2017

Winstock AR, Mitcheson LR, Deluca P et al (2011a) Mephedrone, new kid for the chop? Addiction 106(1):154–161. https://doi.org/10.1111/j.1360-0443.2010.03130.x

Winstock A, Mitcheson L, Ramsey J et al (2011b) Mephedrone: use, subjective effects and health risks. Addiction 106(11):1991–1996. https://doi.org/10.1111/j.1360-0443.2011.03502.x

Winstock AR, Barrett M, Ferris J, Maier L (2016) The Global Drug Survey 2016: what we learned from GDS2016—an overview of our key findings. https://www.globaldrugsurvey.com/wp-content/uploads/2016/06/TASTER-KEY-FINDINGS-FROM-GDS2016.pdf. Accessed 13 Oct 2017

Winstock A, Barratt M, Ferris J, Maier L (2017) Global Drug Survey 2017: global overview and highlights, 24 May. Global Drug Survey, London. https://www.globaldrugsurvey.com/wp-content/themes/globaldrugsurvey/results/GDS2017_key-findings-report_final.pdf. Accessed 13 Oct 2017

Wood DM, Hunter L, Measham F, Dargan PI (2012) Limited use of novel psychoactive substances in South London nightclubs. QJM 105(10):959–964. https://doi.org/10.1093/qjmed/hcs107

Chapter 10
The Effects and Risks Associated with Synthetic Cathinones Use in Humans

Laurent Karila and Amine Benyamina

Abstract New psychoactive substances (NPS) have drastically modified the world drug scene. Synthetic or substituted cathinones (SCs), a popular class of NPS, are β-keto amphetamine analogues. These drugs are also known as legal highs, research chemicals, bath salts, plant food, or glass cleaner, and labeled «not for human use» or «not tested for hazards or toxicity» . However, there is a lack of epidemiological data concerning the SCs. Several factors have contributed to their increasing popularity, including their falsely legal image, their more reasonable costs, and their distribution based on the new technologies. SCs are mainly used for their stimulant properties, often serving as a replacement for others illicit stimulant drugs. The psychoactive and sympathomimetic effects of these drugs resemble those of amphetamine, methamphetamine, ecstasy, and cocaine, albeit to varying degrees. Clinical effects of SCs are individual-, dose- and route of administration-dependent. The primary effects sought by the users include euphoria, empathy, increased sexual performance, and increased sociability. Recent reports on abuse of novel synthetic cathinone derivatives call attention to the serious physical and psychological risks resulting from their consumption, thereby emphasizing the growing use of these drugs might constitute an important public health issue. Acute toxicity is the leading cause of SCs-induced fatalities. Users report a number of negative physical and psychiatric effects associated with SCs use. Cardiac, psychiatric, and neurological adverse effects are the most common reported ones requiring medical care. SCs use may also lead to violence, homicidal combative behavior, self-mutilation, coma, and death.

L. Karila (✉)
Addiction Research and Treatment Center, Paul Brousse Hospital, Paris-Sud University, INSERM U1000, 12 Avenue Paul Vaillant Couturier, Villejuif 94800, France
e-mail: laurent.karila@aphp.fr

A. Benyamina
Addiction Research and Treatment Center, Paul Brousse Hospital (AP-HP), Paris-Sud University, Villejuif 94800, France

A. Benyamina
CESP, Paris, France

J. B. Zawilska (ed.), *Synthetic Cathinones*, Current Topics in Neurotoxicity 12,
https://doi.org/10.1007/978-3-319-78707-7_10

Keywords New psychoactive substances · Synthetic cathinones Substituted cathinones · Substance abuse · Substance use disorder Mephedrone · Methylone · MDPV · α-PVP · 4-MEC · Adverse effects Fatalities

10.1 Introduction

New psychoactive substances (NPS) have modified the world drug scene (EMCDDA 2017) and the current landscape of addiction. Substituted or synthetic cathinones (SCs) ("legal highs"), a popular group of NPS, are often produced and used to mimic the effects of controlled drugs such as cocaine, MDMA, ecstasy, and methamphetamine. These β-keto amphetamine analogues are also known as research chemicals, bath salts, plant food or glass cleaner and labeled «not for human use» or «not tested for hazards or toxicity» (Cottencin et al. 2014).

Mephedrone (4-methylmethcathinone), methylone (3,4-methylenedioxymethcathinone), 3,4-methylenedioxypyrovalerone (MDPV), 4-methylethcathinone (4-MEC), 3-fluoromethcathinone (3-FMC), 4-fluoromethcathinone (4-FMC), buphedrone (α-methylaminobutyrophenone), butylone (β-keto-*N*-methyl-3,4-benzodioxyolybutanamine), methedrone (4-methoxymethcathinone), naphyrone (naphthylpyrovalerone), pentedrone (α-methylaminovalerophenone) are examples of the most well-known SCs (Karila and Reynaud 2011). The five most commonly seized SCs in 2015 were: α-pyrrolidinovalerophenone (α-PVP), 3-methylmethcathinone (3-MMC), ethylone. 4-chloromethcathinone (4-CMC), and pentedrone (EMCDDA 2017). The overwhelming majority of these drugs is produced in China and Southeast Asian countries (Cottencin et al. 2014) but also in Europe (Poland, Netherlands) (EMCDDA 2017).

The Internet has emerged as the new marketplace for NPS, playing a major role in providing information on acquisition, synthesis, extraction, identification, and substance use. Manufacturers, suppliers, retailers, website-hosting, and payment processing services may all be based in different countries (Randolph 2014), making this drug business particularly difficult to be controlled. Constant modification of the chemical structure by covert laboratories allows manufacturers to stay one step ahead of the legal process. All these compounds are also intentionally mislabeled and sold online under slang terms.

At the end of 2015, a total of 644 NPS were registered to the United Nations Office of Drugs and Crime (UNODC) Early Warning Advisory on NPS by 102 different countries (UNODC 2016). The rapid spread of NPS forces member countries of the European Union to adapt their response to the potential new dangers that may cause (EMCDDA 2010).

Recent reports on the SCs abuse call attention to the serious physical and psychological risks resulting from their use, thereby emphasizing the growing use of these drugs might constitute an important public health issue. There is a lack of epidemiological data concerning the new psychoactive substances. The main sources providing information for the study of NPS are the European Monitoring Center for Drugs and Drug Abuse (EMCDDA) (reporting form, risk assessment report, drug profile), the European Union Early Warning System (EU-EWS) reports, the ESPAD School Survey Project, and the National Reitox reports as well as official governmental and international organizations documents, the Internet underground and governmental websites and the discussion groups (EMCDDA 2014a).

In July 2012, synthetic cathinones and derivatives were classified as illicit substances by the Agence Nationale de Sécurité du Médicament (ANSM) in France (Journal Officiel 2012). The Drug Enforcement Administration (DEA) places 10 synthetic cathinones [4-MEC; 4-methyl-α-pyrrolidinopropiophenone (4-MePPP); α-PVP; butylone (bk-MBDB); pentylone (bk-MBDP); 4-FMC; 3-FMC; naphyrone; α-pyrrolidinobutiophenone (α-PBP)] and their optical, positional, and geometric isomers, salts and salts of isomers, into schedule I of the Controlled Substances Act (DEA 2017). In the United Kingdom, MDPV was first classified as a Class B drug under The Misuse of Drugs Act 1971 (Amendment) Order 2010 and has been now placed under the New Psychoactive Act 2016 (Psychoactive Substance Act 2016). In March 2015, MDPV was included in Schedule II of the Convention on Psychotropic Substances of 1971 by the UNODC (2015).

In this chapter, we will review the available data regarding the use and effects of SCs in humans in order to highlight their impact on public health. We will focus on four compounds as examples: mephedrone, MDPV, methylone, and α-PVP. To reach this objective, a literature search was performed on two representative databases (PubMed, Google Scholar), the Erowid Center website (a US non-profit educational organization that provides information about psychoactive plants and chemicals), and various governmental websites. The terms used for the database search were: "new psychoactive substances", "synthetic cathinones", "substituted cathinones", "substance abuse", "substance use disorder", "mephedrone", "methylone", "MDPV", "α-PVP", "adverse effects", and "fatalities". The search was limited to years 2005–2017. The literature search conducted led to the identification of 56 potentially relevant articles. All articles were screened from their abstracts to determine their relevance in the framework of the current review.

10.2 History of Synthetic Cathinones

Khat was used by Egyptians in ancient times for its stimulating psychoactive properties during mystical ceremonies to bring man to a divine rank. After the appearance of Islam in the seventh century AD, the Sufis of the Arabian Peninsula also used it to "get closer to God" and "intensify their mystical experiences". The first scientific traces are mentioned at the beginning of the tenth century AD by Al

Biruni, a Persian scientist and biologist, who describes in his encyclopedia "Kitab al-Saidala fi al-Tibb" this plant imported from Turkestan and used in liver diseases and biliary. By the thirteenth and fourteenth centuries, khat was consumed in the south of the Arabian Peninsula by warriors to fight fatigue and have courage, and by merchants to endure boredom. Other sources even report that the use of khat in Ethiopia and Yemen was prior to the discovery of coffee. *Catha edulis* was discovered in Yemen by an eighteenth-century botanist called Peter Forskal (Krikorian 1984).

The main route of khat consumption is oral. About 100–200 g of fresh leaves are chewed in the Arabian Peninsula and in certain regions of Eastern Africa due to their stimulant effects (Griffiths et al. 2010; see also Chap. 3). The juice that comes out is swallowed immediately. The taste is bitter and astringent. The residues are kept against the cheek for several hours before being spit out. The chewing of the leaves gives off an aromatic odor, and causes a brownish coloration of the teeth and greenishness of the tongue due to the presence of tannins. The acidity of the leaves leads users to accompany their chewing with soda and cigarettes in large quantities. Cathinone is the stimulant alkaloid found in the leaves of the khat bush. It was found to be chemically similar to ephedrine, cathine, and other amphetamines. It produces amphetamine-like psychostimulatory effects, including euphoria, alertness, and psychomotor hyperactivity, but its potency appears to be lower than that of amphetamine (Zawilska and Wojcieszak 2013; see also Chap. 3).

Many designer drugs were first synthesized by chemists in the pharmaceutical industry or in academia in a search for potential new pharmaceuticals (Griffiths et al. 2010). Methcathinone, synthesized in the first half of the twentieth century (Sanchez 1929), was the first synthetic cathinone with reports of abuse beginning in the early 90s (Emerson and Cisek 1993). Mephedrone was first described in 1929 in the *Bulletin de la Société Chimique de France* (Sanchez 1929). Butylone, also known as bk-MBDB, was first synthesized by Koeppe and collaborators in 1967. It has been available at least since 1994, but its popularity on the synthetic drug market remains marginal (Meyer et al. 2010) In 1969, MDPV was initially developed as a treatment for chronic fatigue and as an appetite suppressant by Boehringer Ingelheim (EMCDDA 2014b). α-PVP was first described in 1963 in the α-pyrrolidino-valerophenones (Wander 1963) and the α-pyrrolidino ketones literature (Thomae 1963). The synthesis of α-PVP was also described in 1967 in a patent on α-pyrrolidino ketones by Boehringer Ingelheim (Boehringer Ingelheim 1967). Methylone (3,4-methylenedioxymethcathinone) was first synthesized in 1996 as a potential antidepressant and anti-Parkinsonian agent (Jacob and Sulgin 1996). It, however, never resulted in a commercialized pharmaceutical product. It was initially described in 2004 as a liquid solution, sold as a vanilla-scented room odorizer (Erowid a). Their abuse was not reported until the early twenty-first century, when they were claimed to be legal alternatives to MDMA on Internet drug websites (Morris 2010).

10.3 Routes of Administration

Synthetic cathinones are most frequently used as white/off-white powder, crystalline mixtures (bath salts), or as tablets (Wood et al. 2012). Users frequently snort powder (via a key or «keying») and/or ingest powder rolled up in cigarette paper («bombing») (cited in Zawilska and Wojcieszak 2013, 2017). The intravenous route or «slam», including for example mephedrone, MDPV, or 4-MEC, represents a major health concern (Aides/Sidaction/AMG/Inserm 2013). Some users mix routes, e.g., intravenous with oral, or nasal with oral. Mixing a synthetic cathinone with heroin in a «speedball» type of use has been reported. In some recreational settings, young drug users could start their injecting career with synthetic cathinones. Ex-cocaine users, long-term abstinent ex-opiate users, drug treatment clients, prisoners, and men who have sex with men seem to be keen to slam cathinones (EMCDDA 2014c). Another route of administration that needs to be addressed consists of vaporizing drugs in e-cigarettes (Marusich et al. 2016). Rectal administration ("plug"), gingival delivery, inhalation, intramuscular injection are less common routes (Prosser and Nelson 2012). There have been occasional reports of «eyeballing» (synthetic cathinones inserted into the eyes) (Info.sheet: Naphyrone). Binge consumption of mephedrone, but also other cathinones, is frequently described, based over a 9 h period and 30 min to 2 h between doses, in social contexts (e.g., friends' homes, rave parties, nightclubs, music festival), and its use is frequently associated with other drugs (e.g., alcohol, cocaine, ecstasy, cannabis) (EMCDDA 2010).

10.4 Clinical Pharmacology

The designer SCs are a part of the larger family of stimulants that increase extracellular levels of the monoamine neurotransmitters dopamine, noradrenaline, and serotonin via facilitation of extracellular release and reuptake inhibition (see Chap. 6). As examples, mephedrone and methylone act not only as nonselective monoamine uptake inhibitors like cocaine but also increase serotonin release similarly to MDMA. MDPV acts as a potent and selective monoamine reuptake blocker but has less impact on monoamine release than cocaine. By analogy to MDPV, α-PVP acts as a monoamine transporter inhibitor and does not induce transporter-mediated efflux (Glennon and Young 2016; Marusich et al. 2014; Meltzer et al. 2006; Zawilska and Wojcieszak 2017).

SCs are often used as an alternative to illicit drugs. They can be attractive substances for individuals seeking stimulant psychoactive effects for recreational purposes or for substance addicted individuals who want to avoid being involved in illegal actions. Other reasons given by the users is that SCs, in comparison to cocaine, MDMA or methamphetamine, are less expensive products (Karila and Reynaud 2010, 2011; see also Chap. 9).

Table 10.1 Examples of mephedrone, MDPV, methylone, and α-PVP pharmacokinetic effects

Mephedrone	MDPV	Methylone	α-PVP
• Used intranasally, it elicits rapid and short-lasting effects which appear within minutes, reaches the peak in less than 30 min, and last for less than an hour • Orally ingested mephedrone (powder and tablet) exerts its psychoactive effects after 45 min–2 h which then last for 2–4 h (Schifano et al. 2011)	• Orally ingested, it exerts its psychoactive effects after 30 min–3 h, which then last for 0.5–3 h • The decrease of the psychoactive effects vary with the administration route from 2–48 h (oral) to 1–7 days (intranasal) (Adamowicz et al. 2013; Baumann et al. 2017)	• Psychoactive effects last from 3–5 h • Onset of desired effects is typically seen within 15–60 min post-oral ingestion • Desired effects last approximately 30–45 min (WHO 2014)	• α-PVP delivers a rapid high that can last 1–3 h • Symptoms last from hours to days with lingering effects for weeks after it has been consumed • The peak is reached in 10 up to 40 min (Stanciu et al. 2017)

Psychotropic effects of SCs are individual-, dose- and route of administration-dependent (Prosser and Nelson 2012). Users may consume multiple doses during a session to prolong the desired effects. Sympathomimetic, subjective, and reinforcing effects are the three main effects of these drugs. The effects generally start about 10–20 min after dosing, and then decrease over 6–12 h (Table 10.1).

In general, positive subjective effects sought by the users include empathy, euphoria, elevated mood, increased self-esteem, openness in communication, talkativeness, increased concentration, increased sociability and capacity to work, increased productivity, increased motivation, alertness, motor excitation, music sensitivity, reduced appetite, insomnia, sexual arousal, and increased libido (Rosenbaum et al. 2012; Stanciu et al. 2017). Desired effects of mephedrone, MDPV, methylone and α-PVP are described in Table 10.2.

10.5 Reported Adverse Effects

Reported adverse effects concern not only SCs acute toxicity but also chronic toxicity, liability toward abuse, and addictive potential. The negative side effects increase with higher doses. Cardiac, neurological and psychiatric adverse effects are the most common reported ones (Cottencin et al. 2014; Karila et al. 2015). In addition to the cathinone effects, contaminants, such as benzocaine, lidocaine, procaine, caffeine, or controlled drugs such as cocaine, amphetamine, ketamine, and

Table 10.2 Desired effects of mephedrone, MDPV, methylone, and α-PVP

Mephedrone	MDPV	Methylone	α-PVP
• Elevated mood • Feelings of intense euphoria, • Increased self-esteem • Tachypsychia • Reduced perception of tiredness • Increased concentration, alertness, talkativeness • Empathy • Disinhibition • A sense of being speed up • Mild sexual stimulation (Karila et al. 2015)	• Empathy • Euphoria • Elevated mood • Increased self-esteem • Talkativeness, increased concentration • Increased sociability • Increased productivity • Increased motivation • Alertness • Motor excitation • Sexual arousal • Increased libido (Stanciu et al. 2017)	• Amphetamine-like stimulation, such as a calm euphoria, happiness, thought acceleration, alertness, restlessness, reduced fatigue, increased locomotor activity • MDMA entactogenic effects (e.g., strong sense of empathy, reduced fear) (Shimizu et al. 2007)	• Enhanced energy • Euphoria • Empathy • Openness • Increased libido (Marusich et al. 2016; Zawilska and Wojcieszak 2017)

piperazine compounds, could play a significant role in adverse effects (Shimizu et al. 2007; Weaver et al. 2015; Zawilska and Wojcieszak 2013).

10.5.1 Physical Adverse Effects

There are numerous physical adverse effects of SCs use (Karila et al. 2015) affecting many systems:

- Cardiovascular effects: tachycardia, hypovolemia, hypertension, palpitations, chest pain, ST segment alterations, myocarditis, cardiac arrest;
- Pulmonary effects: shortness of breath, tachypnea, respiratory failure and arrest, respiratory acidosis;
- Renal effects: increased serum creatinine, kidney damage, acute renal failure, hyponatremia, hyperkalemia, hyperuricemia;
- Gastrointestinal and hepatic effects: emesis, nausea, abdominal pain, abnormal liver function tests, liver failure;
- Central nervous system effects: headache, insomnia, dizziness, seizures, tremor, confusion, collapse, confusion, dizziness, drowsiness, dystonia, headache, hyperreflexia, myoclonus, paresthesias;
- Hematologic effects: disseminated intravascular coagulation, thrombocytopenia, anemia;
- Musculoskeletal effects: elevated creatinine kinase, rhabdomyolysis, compartment syndrome, peripheral vasoconstriction;
- Ophthalmic effects: mydriasis, blurred vision, nystagmus;
- ENT effects: epistaxis, oral and pharyngeal effects, trismus, bruxism;
- Infectious diseases: HIV, hepatitis B and C, sexually transmitted diseases (especially when drugs are consumed during "Chem Sex Parties");
- Other effects: hyperthermia, skin rash, diaphoresis, bone pain, necrotizing fasciitis, serotonin syndrome;
- Consequences of intravenous route use: vein blockage, skin erosion, local infection, abscess, scab, lump, gangrenous tissue, blood clots and large holes at overused injecting sites.

10.5.2 Psychiatric Adverse Effects

Psychiatric adverse effects are agitation, aggression, anxiety, paranoia, paranoid delusion, depersonalization/derealization, panic attacks, anorexia, visual and auditory hallucinations, psychosis, anhedonia, depression, suicidal thoughts/actions, self-stabbings, repeated self-mutilation and risk behavior without evidence of psychosis or depression comorbidity, cognitive disorders (long-term cognitive impairments, disorientation to names, place and time, loosening of association) (Karila et al. 2015, 2017).

10.6 Fatalities

SCs have been implicated in several fatalities across the world. Case reports on synthetic cathinones-related fatalities are described in the literature. By 2012, 128 mephedrone-associated fatalities have been reported (Schifano et al. 2012). Studies have not determined the minimum lethal dose of mephedrone (EMCDDA 2010). A number of fatal intoxications involving methylone use have been reported (Karila et al. 2015). In 2014, in Europe, a total of 99 fatalities associated with MDPV were analytically confirmed. These were reported by Finland (40), the United Kingdom (UK) (32), Sweden (21), Poland (3), France, Austria, Norway (one for each country) (EMCDDA 2014b). From 2012 to 2015, a total of 115 deaths—in which α-PVP was analytically confirmed—were reported by eight Member States in Europe. In 23 cases, the drug was reported as the cause of death or as a contributing factor. In five cases, α-PVP was the only substance detected. The cause of death varied and included sudden cardiac arrest, suicides, or accidents (Karila et al. 2017; Zawilska and Wojcieszak 2017).

10.7 Addictive Potential

Repeated dosing of SCs likely leads to tolerance. Development of physiological dependence among regular users, which is manifested by tolerance and a withdrawal syndrome, is reported (Weinstein et al. 2017). The addictive potential of mephedrone exists (Karila and Reynaud 2011; Winder et al. 2013). Mephedrone taken intranasally could have a potential for abuse comparable to that of methamphetamine (Karila and Reynaud 2011). MDPV has reinforcing properties and activates the brain reward circuitry, suggesting a potential for addiction in humans (Gannon et al. 2017). The development of craving has been reported after frequent use of high doses of MDPV (Andrabi et al. 2015) Methylone is also highly susceptible to have an addictive potential, probably comparable to the one of MDMA (Erowid b). α-PVP involves high risk for abuse in human users (Aarde et al. 2015). This drug seems to possess a potential for compulsive abuse similar to that of methamphetamine. Self-reported experiences from users include extreme craving and risk of binge consumption (Cheong et al. 2017). However, further studies are needed to identify the addictive potential of SCs in humans.

References

Aarde SM, Creehan KM, Vandewater SA et al (2015) In vivo potency and efficacy of the novel cathinone alpha-pyrrolidinopentiophenone and 3,4-methylenedioxypyrovalerone: self-administration and locomotor stimulation in male rats. Psychopharmacology 232 (16):3045–3055. https://doi.org/10.1007/s00213-015-3944-8

Adamowicz P, Gil D, Skulska A, Tokarczyk B (2013) Analysis of MDPV in blood—determination and interpretation. J Anal Toxicol 37(5):308–312. https://doi.org/10.1093/jat/bkt025

AIDES/Sidaction/AMG/Inserm (2013) SLAM, première enquête qualitative en France. Février 2013. https://issuu.com/ministere-sante/docs/rapport_morlat_2013_mise_en_ligne

Andrabi S, Greene S, Moukaddam N, Li B (2015) New drugs of abuse and withdrawal syndromes. Emerg Med Clin North Am 33(4):779–795. https://doi.org/10.1016/j.emc.2015.07.006

Baumann MH, Bukhari MO, Lehner KR et al (2017) Neuropharmacology of 3,4-methylenedioxypyrovalerone (MDPV), its metabolites, and related analogs. Curr Top Behav Neurosci 32:93–117. https://doi.org/10.1007/7854_2016_53

Boehringer Ingelheim G (1967) Pyrrolidino ketones. Unit 3314970 A

Cheong JH, Choi MJ, Jang CG et al (2017) Behavioral evidence for the abuse potential of the novel synthetic cathinone alpha-pyrrolidinopentiothiophenone (PVT) in rodents. Psychopharmacology 234(5):857–867. https://doi.org/10.1007/s00213-017-4526-8

Cottencin O, Rolland B, Karila L (2014) New designer drugs (synthetic cannabinoids and synthetic cathinones): review of literature. Curr Pharm Des 20(25):4106–4111

DEA (2017) Schedules of controlled substances: placement of 10 synthetic cathinones into schedule I. https://www.federalregister.gov/documents/2017/03/01/2017-03974/schedules-of-controlled-substances-placement-of-10-synthetic-cathinones-into-schedule-i. Accessed May 2017

EMCDDA (2014a) European monitoring centre for drugs and drug addiction. European drug report 2014: trends and developments. Publications Office of the European Union 2014, Luxembourg, 80 pp

EMCDDA (2014b) EMCDDA–Europol joint report on a new psychoactive substance: MDPV (3,4-methylenedioxypyrovalerone). Publications Office of the European Union, Luxembourg

EMCDDA (2014c) Perspectives on drugs. Injection of synthetic cathinones. http://www.emcdda.europa.eu/topics/pods/synthetic-cathinones-injection_en. Accessed 5 May 2014

EMCDDA (2017) European drug report 2017: trends and developments. Publications Office of the European Union, Luxembourg

Emerson T, Cisek J (1993) Methcathinone: a Russian designer amphetamine infiltrates the rural midwest. Ann Emerg Med 22:1897–1903

Erowid (a) http://www.erowid.org/chemicals/methylone/methylone_info1.shtml. Accessed July 2010

Erowid (b) Methylone addictive potential. https://www.erowid.org/experiences/exp.php?ID=27823

European Monitoring Centre for Drugs and Drug Addiction (EMCDDA) (2010) Risk assessment report of a new psychoactive substance: 4-methylmethcathinone (mephedrone). http://www.emcdda.europa.eu/system/files/publications/559/2010_Mephedrone_Joint_report_279

Gannon BM, Rice KC, Collins GT (2017) Reinforcing effects of abused 'bath salts' constituents 3,4-methylenedioxypyrovalerone and alpha-pyrrolidinopentiophenone and their enantiomers. Behav Pharmacol 28(7):578–581. https://doi.org/10.1097/FBP.0000000000003152017

Glennon RA, Young R (2016) Neurobiology of 3,4-methylenedioxypyrovalerone (MDPV) and α-pyrrolidinovalerophenone (α-PVP). Brain Res Bull 126(Pt 1):111–126. https://doi.org/10.1016/j.brainresbull.2016.04.011

Griffiths P, Lopez D, Sedefov R et al (2010) Khat use and monitoring drug use in Europe: the current situation and issues for the future. J Ethnopharmacol 132(3):578–583. https://doi.org/10.1016/j.jep.2010.04.046

Info.sheet: Naphyrone. http://crew2000.org.uk. Accessed 8 Nov 2012

Jacob P, Shulgin A (1996) Inventors, Neurobiological Technologies, Inc., assignee. Preparation of novel N-substituted-2-amino-3′,4′-methylenedioxypropiophenones as anti-depressant and anti-parkinsonism agents. US Patent WO9639133, 6 June 1996

Journal Officiel (2012) Arrêté du 27 juillet 2012 modifiant les arrêtés du 22 février 1990 fixant la liste des substances classées comme stupéfiants et la liste des substances psychotropes. https://www.legifrance.gouv.fr/affichTexte.do?cidTexte=JORFTEXT000026246525&categorieLien=id. Accessed 8 March 2016

Karila L, Reynaud M (2010) Mephedrone: a designer drug legally available on the Web. Presse Med 39(7–8):834–835

Karila L, Reynaud M (2011) GHB and synthetic cathinones: clinical effects and potential consequences. Drug Test Anal 3(9):552–559. https://doi.org/10.1002/dta.210

Karila L, Megarbane B, Cottencin O, Lejoyeux M (2015) Synthetic cathinones: a new public health problem. Curr Neuropharmacol 13(1):12–20. https://doi.org/10.2174/1570159X13666141210224137

Karila L, Lafaye G, Scocard A et al (2017) MDPV and alpha-PVP use in humans: the twisted sisters. Neuropharmacology. https://doi.org/10.1016/j.neuropharm.2017.10.007

Krikorian AD (1984) Kat and its use: an historical perspective. J Ethnopharmacol 12(2):115–178

Marusich JA, Antonazzo KR, Wiley JL et al (2014) Pharmacology of novel synthetic stimulants structurally related to the "bath salts" constituent 3,4-methylenedioxypyrovalerone (MDPV). Neuropharmacology 87:206–213. https://doi.org/10.1016/j.neuropharm.2014.02.016

Marusich JA, Lefever TW, Blough BE et al (2016) Pharmacological effects of methamphetamine and alpha-PVP vapor and injection. Neurotoxicology 55:83–91. https://doi.org/10.1016/j.neuro.2016.05.015

Meltzer P, Butler D, Deschamps J, Madras B (2006) 1-(4-Methylphenyl)-2-pyrrolidin-1-yl-pentan-1-one (Pyrovalerone) analogues: a promising class of monoamine uptake inhibitors. J Med Chem 49(4):1420–1432. https://doi.org/10.1021/jm050797a

Meyer MR, Wilhelm J, Peters FT, Maurer HH (2010) Beta-keto amphetamines: studies on the metabolism of the designer drug mephedrone and toxicological detection of mephedrone, butylone, and methylone in urine using gas chromatography-mass spectrometry. Anal Bioanal Chem 397(3):1225–1233. https://doi.org/10.1007/s00216-010-3636-5

Morris H (2010) Mephedrone: the phantom menace. Vice:98–100

Prosser JM, Nelson LS (2012) The toxicology of bath salts: a review of synthetic cathinones. J Med Toxicol 8(1):33–42. https://doi.org/10.1007/s13181-011-0193-z

Psychoactive Substances Act (2016). http://www.legislation.gov.uk/ukpga/2016/2/contents/enacted. Accessed 29 May 2016

Randolph SA (2014) Synthetic drugs: bath salts and spice. Workplace Health Saf 62(2):88

Rosenbaum CD, Carreiro SP, Babu KM (2012) Here today, gone tomorrow…and back again? A review of herbal marijuana alternatives (K2, Spice), synthetic cathinones (bath salts), kratom, Salvia divinorum, methoxetamine, and piperazines. J Med Toxicol 8(1):15–32. https://doi.org/10.1007/s13181-011-0202-2

Sanchez S (1929) Sur un homologue de l'ephedrine. Bull Soc Chim Fr 45:284–286

Schifano F, Albanese A, Fergus S et al (2011) Mephedrone (4-methylmethcathinone; 'meow meow'): chemical, pharmacological and clinical issues. Psychopharmacology 214(3):593–602. https://doi.org/10.1007/s00213-010-2070-x

Schifano F, Corkery J, Ghodse A (2012) Suspected and confirmed fatalities associated with mephedrone (4-methylmethcathinone, "meow meow") in the United Kingdom. J Clin Psychopharmacol 32:710–714. https://doi.org/10.1097/JCP.0b013e318266c70c

Shimizu E, Watanabe H, Kojima T et al (2007) Combined intoxication with methylone and 5-MeO-MIPT. Prog Neuropsychopharmacol Biol Psychiatry 31(1):288–291

Stanciu CN, Penders TM, Gnanasegaram SA et al (2017) The behavioral profile of methylenedioxypyrovalerone (MDPV) and α-pyrrolidinopentiophenone (PVP)—A systematic review. Curr Drug Abuse Rev. https://doi.org/10.2174/1874473710666170321122226

Thomae K (1963) α-Pyrrolidinoketones, patent specification 933507

United Nations Office on Drugs and Crime (UNODC) (2015) Decision 58/12—Inclusion of 3,4-methylenedioxypyrovalerone (MDPV) in Schedule II of the Convention on Psychotropic Substances of 1971.74. https://www.unodc.org/documents/commissions/CND/CND_Sessions/CND_58/2015_Desicions/Desicion_58_12.pdf

UNODC (2016) World drug report 2016. United Nations Publication, Sales No. E.16.XI.7

Wander A (1963) α-Pyrrolidinovalerophenones, patent specification 927475

Weaver MF, Hopper JA, Gunderson EW (2015) Designer drugs 2015: assessment and management. Addict Sci Clin Pract 10:8. https://doi.org/10.1186/s13722-015-0024-7

Weinstein AM, Rosca P, Fattore L, London ED (2017) Synthetic cathinone and cannabinoid designer drugs pose a major risk for public health. Front Psychiatry 8:156. https://doi.org/10.3389/fpsyt.2017.00156

WHO (2014) Methylone (bk-MDMA)—critical review report—expert committee on drug dependence thirty-sixth meeting—Geneva, 16–20 June 2014

Winder GS, Stern N, Hosanagar A (2013) Are "bath salts" the next generation of stimulant abuse? J Subst Abuse Treat 44(1):42–45. https://doi.org/10.1016/j.jsat.2012.02.003

Wood D, Hunter L, Measham F, Dargan P (2012) Limited use of novel psychoactive substances in South London nightclubs. QJM 105:959–964

Zawilska JB, Wojcieszak J (2013) Designer cathinones—an emerging class of novel recreational drugs. Forensic Sci Int 231(1–3):425–453. https://doi.org/10.1016/j.forsciint.2013.04.015

Zawilska J, Wojcieszak J (2017) α-Pyrrolidinophenones: a new wave of designer cathinones. Forensic Toxicol 35(2):201–216. https://doi.org/10.1007/s11419-016-0353-6

Chapter 11
Concluding Remarks: Where We Are and Where Do We Go from Here?

Jolanta B. Zawilska

Abstract The ever-changing world of synthetic cathinones constitutes a major challenge to the provision of current and reliable scientific knowledge on these substances. This chapter briefly outlines major issues and problems related to synthetic cathinones that are discussed in the present interdisciplinary book. It also points to future efforts that are critically needed to develop new effective prevention and intervention strategies aimed at reduction of negative health, social and economic consequences related to the use of cathinones.

Keywords Novel psychoactive substances · Synthetic cathinones
Harm reduction strategies

Over the last decade, the popularity and use of novel psychoactive substances (NPS) have increased dramatically, resulting in a global drug abuse crisis. In their early years, NPS could be purchased through a range of legal retail outlets. However, over the years more sales have gone through the Internet, including the Darkweb, largely in response to legal classification. Synthetic cathinones represent the second largest group of NPS. Almost 130 cathinones were reported to the European Monitoring Centre for Drugs and Drug Addiction, and new compounds from this class are likely to appear on the "recreational" drug market. The nature and extent of NPS (including cathinones) use remain largely unknown, as does whether its growth can be slowed.

Synthetic cathinones are chemically related to cathinone, a naturally occurring stimulant found in the leaves and stalks of the plant *Catha edulis* Forsk (Chap. 3). They exert effects similar to those of common illicit stimulant drugs such as amphetamine, methamphetamine, cocaine, and MDMA. Due to the actions on the central nervous system, the first compounds from this class were synthesized for medical purposes in the early twentieth century, but began to attract wider attention

J. B. Zawilska (✉)
Department of Pharmacodynamics, Medical University of Łódź,
Muszynskiego 1, 90-151 Łódź, Poland
e-mail: jolanta.zawilska@umed.lodz.pl

J. B. Zawilska (ed.), *Synthetic Cathinones*, Current Topics in Neurotoxicity 12,
https://doi.org/10.1007/978-3-319-78707-7_11

around 2000. The two pioneering representatives of designer cathinones that emerged onto the recreational drug market were methcathinone and 4-methylmethcathinone (mephedrone). Their appearance coincided with the time when the availability and purity of psychostimulant narcotics in Europe were falling and prices were going up. They were soon followed by new compounds, e.g., methylone (3,4-methylenedioxy-*N*-methylcathinone) and 3,4-methylenedioxypyrovalerone (MDPV). Immediately after this first generation become illegal, the so-called second generation emerged as replacements, and when these become controlled in 2010s, cathinones from the third generation began to conquer the market. Alarming data suggests that, by analogy to mephedrone, other synthetic cathinones have joined the ranks of psychostimulatory drugs following an initial period of replacing and displacing ecstasy and cocaine (see Chap. 9).

Despite considerable efforts, our knowledge about the prevalence and motivation of cathinone use, which is crucial for the development of effective and appropriately tailored harm-reduction techniques and initiatives, is still very limited. As emphasized by Cockery and co-workers in Chap. 9, survey instruments and tools need to be improved and international cooperation enhanced to face this challenge.

The ever-changing world of synthetic cathinones is closely related to the difficulties concerned with identifying them and their metabolites (see Chaps. 4 and 5). Most of these compounds are not covered by standard drug testing, meaning that many intoxication cases are probably missed. To detect the increasing range of NPS, including synthetic cathinones, rapid updating of the analytical methods employed in clinical and forensic toxicological laboratories is required.

Although the molecular mechanisms of action of several cathinones have been revealed in in vitro studies (see Chap. 6), in vivo tests are required to determine their potency and efficacy. Understanding the potential characteristics and properties of new molecules, including their effects on brain neurotransmission and behaviour, particularly their rewarding and reinforcing activity (see Chaps. 7 and 8), is another important goal to achieve.

As described in detail in Chap. 10, synthetic cathinones exert numerous toxic effects in humans, some of them fatal. Dealing with the medical complications associated with cathinone use and overdose has become a growing problem for emergency departments and intensive care units worldwide. In order to efficiently address this problem, deeper knowledge about the common doses used, frequency of dosing, and duration of action is needed, as well as of the interaction of cathinones with other drugs, and dependence/addition. Greater effort is also needed to understand why users are looking for new synthetic cathinones, and what they are actually taking. To achieve this goal, interdisciplinary approaches should be implemented.

Last but not least, our success in the fight against various NPS demands the coordination of the united efforts of a multitude of people including physicians, toxicologists, psychologists, analytical chemists, teachers, parents, policeman and lawmakers, as well as national and international agencies and warning systems.

Index

J. B. Zawilska (ed.), *Synthetic Cathinones*, Current Topics in Neurotoxicity 12,
https://doi.org/10.1007/978-3-319-78707-7

The manufacturer's authorised representative in the EU is Springer
Nature Customer Service Centre GmbH, Europaplatz 3, 69115 Heidelberg,
Germany. If you have any concerns regarding our products, please
contact ProductSafety@springernature.com

Printed and bound by CPI Group (UK) Ltd, Croydon, CR0 4YY
30/09/2026
02225655-0003